©2008 C3 Publishing Co.
Published in 2008 by C3 Publishing Co.
72-9 Gonghang Gangseo Seoul 157-812 Korea
Tel. +82 2 2661 2811   Fax. +82 2 2661 2456
E-mail. info@c3-d.com   http://www.c3-d.com

발행처   건축과환경
출판등록   1991년 5월 22일 제13-295호

Publisher   JaeHong Lee 이재홍
Editor   Uje Lee 이우재
Managing Editor   JaeWon Lee 이재원
Editorial Designer   MinJung Kim 김민정
Color Editor   HyunJung Han 한현정
International Marketing   SangMin Lee 이상민
Printed by   Orange Printech co. Ltd. 주.오렌지프린텍
Cover Image   Beijing Capital Museum ©Didier Boy de la Tour

All rights reserved. No part of this publication may be reproduced or transmitted
in any form or by any means, electronic or mechanical, including photocopy,
recording or otherwise, without the prior permission of C3 Publishing Co..
이 책의 저작권은 건축과환경사에 있으며, 저작권법으로 보호됩니다. 광선자 매재 수록을 포함한
어떠한 방법이나 형태로든 무단 복제 혹은 전재를 금합니다.

ISBN   978-89-86780-73-4

Printed in Korea

museum & exhibition space

6	**Beijing Capital Museum** _ AREP 베이징 시립 박물관 _ AREP
20	**ARoS** Aarhus Museum of Modern Art _ Schmidt, Hammer & Lassen 아로스 현대미술관 _ 스미트, 하머 앤 레슨
32	**Palmach Museum of History** _ Zvi Hecker 팔마치 역사 박물관 _ 쯔비 헥커
50	**Archaeological Museum, Saint-Romain-en-Ga** _ Chaix & Morel 생 로맹 엘 갈 고고학 박물관 _ 쉑스 앤 모렐
72	**Jean Tinguely Museum** _ Mario Botta 장 팅겔리 박물관 _ 마리오 보타
82	**MuseumQuartier Wien** _ Ortner & Ortner Architecture 빈 뮤지엄콰르티에 2제 _ 오트너 앤 오트너 아키텍쳐
108	**Miho Museum** _ I. M. Pei 미호미술관 _ I. M. 페이
132	**Lille Fine Arts Museum** _ Jean-Marc Ibos & Myrto Vitart 릴 미술관 _ 이보스 앤 비따르
148	**New Wing of the Van Gogh Museum** _ Kishow Kurokawa 반 고흐 미술관 신관 _ 쿠로카와 키쇼
156	**Wolfsburg Art Museum** _ Schweger + Partner 볼프스부르크 미술관 _ 슈베거 앤 파트너
168	**Museum Sammlung Essl** _ Heinz Tesar 삼믈룽예셀 미술관 _ 하인쯔 테자
176	**Australian Centre for Contemporary Art** _ Wood Marsh Architecture 호주 현대예술센터 _ 우드 마쉬 아키텍쳐

182 Siida _ Juhani Pallasmaa

192 Museum of the Iron _ Bianchini & Lusiardi associates

200 Dynamic Earth _ Michael Hopkins

206 Hedge Building _ Atelier Kempe Thill

214 Inventaariokamari C74, Suomenlinna _ Arkkitehtitoimisto Laiho-Pulkkinen-Raunio

224 Rehabilitation of the Charles V Palace in the Alhambra _ Juan Pablo Rodriguez Frade

234 Invisible Bos _ GAD

238 Gagosian Gallery _ Gluckman Mayner Architects

242 Wind Nest the Finnish-Pavillion _ SARC Architects

250 P. S. 1 Dunscape _ SHoP

258 Turkish Pavilion Hannover Expo 2000 _ Tabanlıoğlu

Beijing Capital Museum
베이징 시립 박물관

AREP
AREP

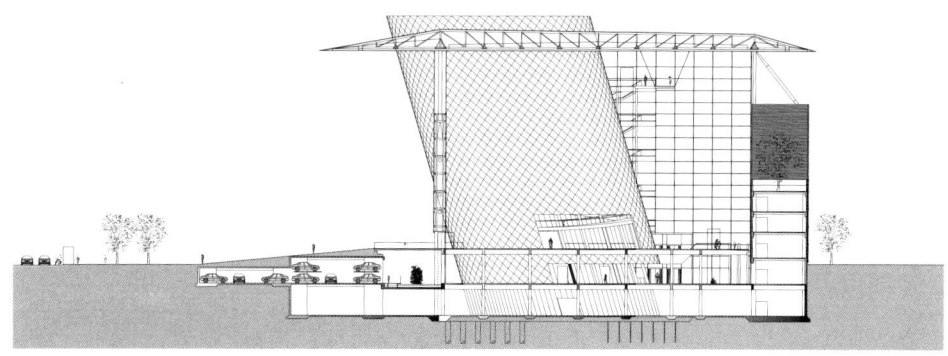
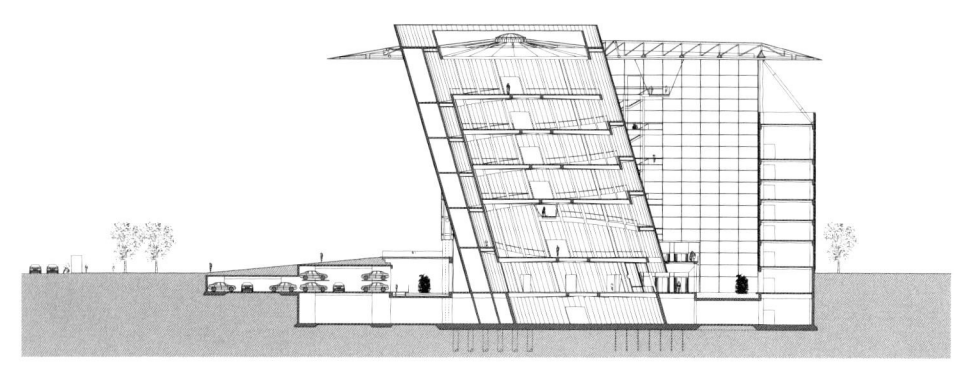
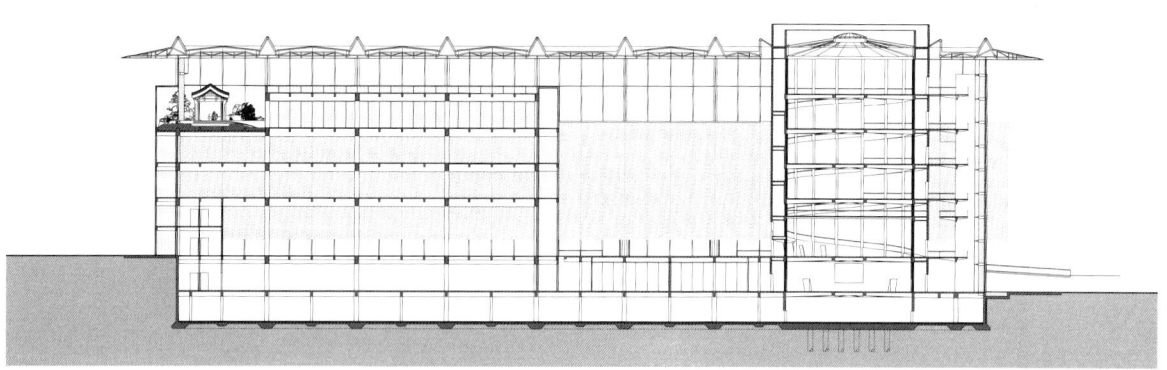
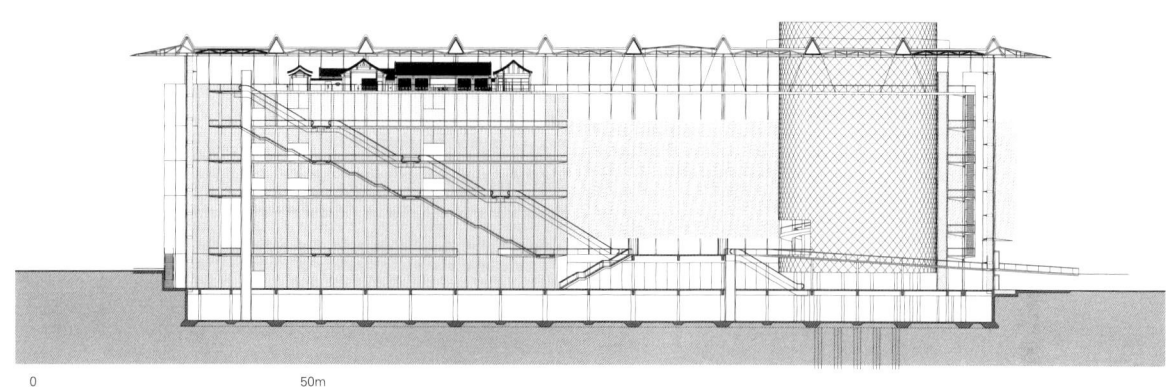

0   50m

8 MUSEUM & EXHIBITION SPACE _ Beijing Capital Museum

Beijing Capital Museum is located on Fu Xing Men Avenue, a major thoroughfare running east-west as a continuation of Xi Chang An Jie Avenue, which runs alongside the Forbidden City and Tien An Men Square. Fu Xing Men Avenue will ultimately be flanked by a series of high-profile cultural amenities including an opera house and a library. The museum was planned as one of the city flagship public building and will participate in the image of Beijing as a worldwide main capital.

The museum has a total surface area of 60,000m^2 protected by a rectangular horizontal roof. It stands back from the avenue in order to create a public square that opens out onto the city and in so doing, showcases the building and creates a court where festivals and cultural events may be held. Underneath this large roof that cantilevers off above the public square, the museum layout is composed like a Chinese palace. Three pavilions surround a central hall on its 3 sides, while, on the fourth one, the main entrance is conceived as the entrance of the Chinese palaces : the visitors slip into the museum under a stone wall which is an allegory of the zigzag path one has to go through to enter. Each of the three objects has its own volume, indoor space and materials specifically related to the symbolic and functional uses they receive.

The first of these objects, the 7,000m^2 'Treasure Gallery', is made of bronze and houses most of the precious objects collection that includes jades, coins, seals, Beijing Opera costumes, paintings etc. It forms a cylinder leaning at an angle. Its unique volume as well as and the way it is inscribed within the main north-facing facade give the entire building its signature identity. It is the object that identifies the museum, mysterious and startling at the same time.

The second volume (40m wide, 35m high and 72m long), with its wooden fascia on both indoor and outdoor surfaces, forms a sort of 'box' which hosts 8,000m^2 of temporary exhibition on the ground floor, 10,000m^2 of permanent collection that retraces the history of Beijing, a part of the collections including porcelain, bronzes, furniture, clothing and religious artifacts, and an educational area containing maps, models and interactive displays that trace the development of the Forbidden City.

The third component is the elongated grey stone block closing off the building on the south side and which contains the media library, the research and museology facilities and areas for museum staff.

The base, the last component of the complex on which the other three volumes are built, contains ancillary museum facilities, including boutiques, restaurants and an auditorium, together with more temporary exhibition space, an underground car park and technical facilities.

The climate and acoustical environment of the extensive indoor areas is tightly controlled to offer visitors maximum comfort. As a result, the museum can host indoor public events during winter and organise performances in the large central atrium. Natural light floods into these extensive spaces through the south and east facing facades and through the roof. It is filtered to minimize glare, shaping the various volumes and bouncing off the different materials.

The programme echoes Beijing's cultural heritage while simultaneously expressing the city's contemporary, forward-looking ethos. The classical attributes of Chinese buildings are reformulated using a contemporary vocabulary to forge a delicate balance between imposed order and freedom of composition, the connection between the monumental and the intimate, and the progressive unfolding of the spaces as visitors move from the outside in.

베이징 시립 박물관은 자금성과 천안문 광장을 지나 장안가와 이어지는 부흥문가에 위치한다. 동서 방향의 간선 도로인 부흥문가에는 오페라 하우스와 도서관을 포함한 일련의 문화시설들이 들어설 예정이다. 이 박물관은 베이징의 주요 공공건물 중 하나로 계획되었으며, 세계적 수도인 베이징의 이미지 강화에 기여할 것이다.

건축면적이 총 6만m^{2}인 박물관은 직사각형 수평 지붕으로 덮여 있으며, 공공광장을 위해 거리에서 안쪽으로 들어가 위치한다. 이로써 도시로 열리면서 건물 자체가 전시되고 축제와 문화 행사가 열리는 마당이 조성된다. 이 공공광장 위에 캔틸레버로 뻗은 커다란 지붕을 가진 박물관의 전체 모습은 중국의 궁전과 유사하다. 세 개의 파빌리온이 세 면에서 중앙홀을 둘러싸고 있으며, 나머지 한 면은 중국 궁전의 입구와 유사한 모습의 주출입구가 들어선다. 방문객들은 석벽 아래에 있는 박물관 안으로 들어가는데, 이 석벽은 지그재그로 된 길을 통해야만 안으로 들어갈 수 있다. 이들 세 매스는 각각 자신의 볼륨과 내부 공간을 가지고 있으며 상징, 기능적 용도와 관련된 재료가 쓰였다.

7천m^2 규모의 첫 번째 매스인 '보물 화랑'은 청동으로 세워졌다. 옥, 엽전, 인장, 전통 오페라 의상, 그림 등과 같은 귀중한 유물을 수장하는 이 매스는 모서리에 의지하는 원통처럼 보인다. 그 독특한 볼륨과 북향의 주 파사드 내에 새겨진 모습은 전체 건물의 상징이 되고, 동시에 박물관에 신비롭고 경이로운 정체성을 부여한다.

내·외부 표면에 목재 띠가 있는 두 번째 볼륨은 '박스'(폭 40m, 높이 35m, 길이 72m) 형상을 띤다. 이곳은 지상층에 8천m^2 규모의 임시 전시장이 있고 그 위에 북경의 역사를 되짚는 도자기, 청동기, 가구, 의복, 제기 등의 유물이 전시되는 1만m^2 규모의 상설 전시장과 자금성의 발전과정을 보여주는 지도, 모형, 인터랙티브 디스플레이 등을 전시하는 교육관이 있다.

세 번째 매스는 건물 남쪽면을 닫는 긴 회색 석조 블록으로 미디어 도서관, 연구 및 박물관학 관련 시설, 그리고 박물관 관리 사무실을 수용한다. 이 세 매스가 들어서는 박물관 단지의 마지막 부분인 기단은 부띠끄, 식당, 강당, 임시 전시 공간, 지하 주차장, 기계실 등 박물관 지원 시설들을 수용한다.

넓은 실내 온도와 음향 환경은 방문객들이 최대한 편안함을 느끼도록 세밀하게 제어된다. 그 결과 겨울에도 실내에서 공공 이벤트를 개최할 수 있고, 대형 중앙 아트리움에서 공연을 기획할 수 있다. 남향 및 동향 파사드와 지붕을 통해 실내로 유입되는 일광은 적절히 여과되어 눈부시지 않으며, 다양하게 형성된 볼륨과 재료들로부터 반사된다.

프로그램은 베이징의 문화적 유산을 반향하면서 동시에 도시의 현대적이고 전향적인 민족성을 표현한다. 중국 건물의 고전적인 속성들이 현대적인 어휘를 통해 재구성되어 강요된 질서와 구성의 자유 사이에서 미묘한 균형을 잡고 기념비성과 친근성을 연계시킨다. 그리고 방문객들이 밖에서 들어올 때 공간들이 하나씩 펼쳐지는 효과를 연출한다.

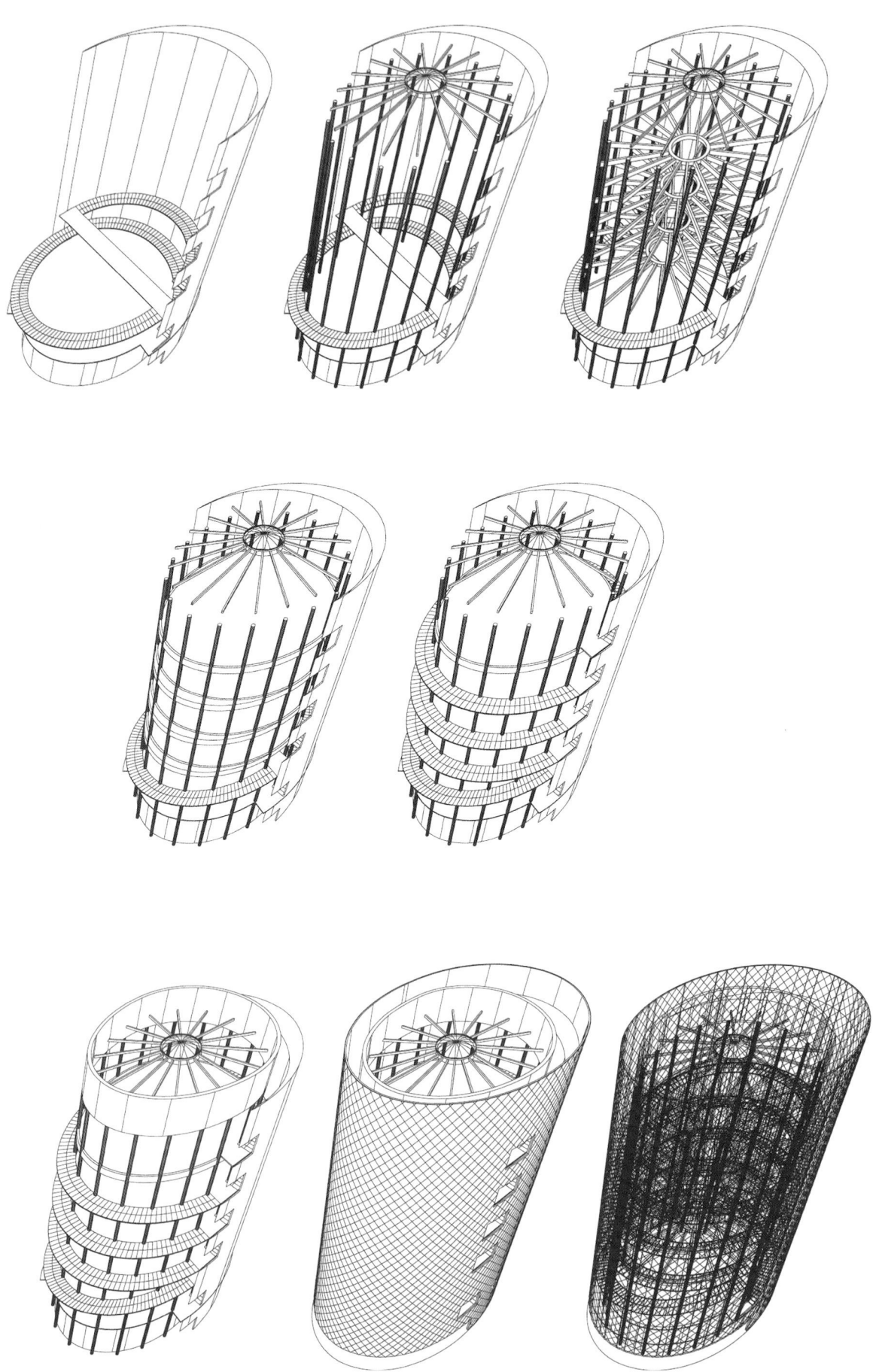

structural details of the bronze cylinder

Prime contractor : Jean-Marie Duthilleul and
Etienne Tricaud (AREF),
Cui Kai (Architecture & Design Institute of the
Ministry of Construction)
Client : City of Beijing
Location : Beijing, China
Photograph : Didier Boy de la Tour-p.9, p.10,
p.13, p.14~15, p.16, p.17, p.18
Tristan Chapuis-p.6~7

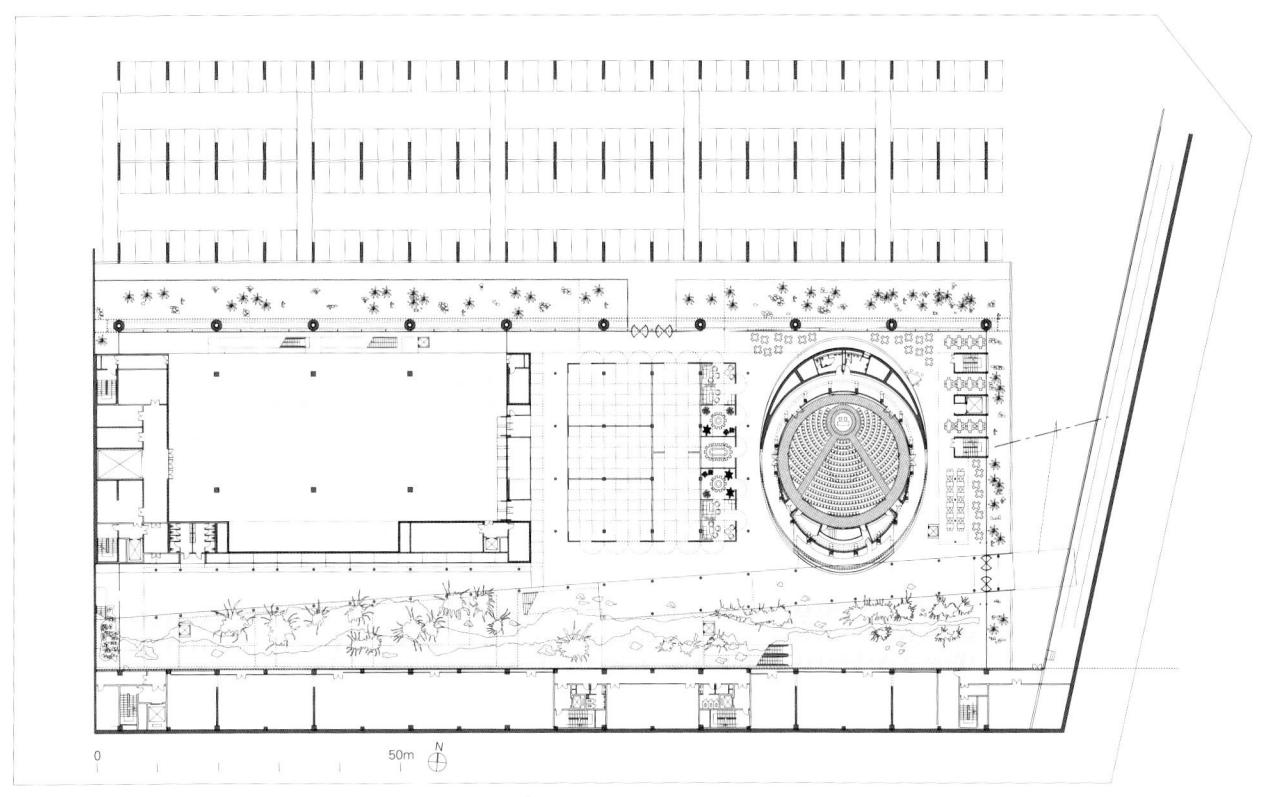

basement

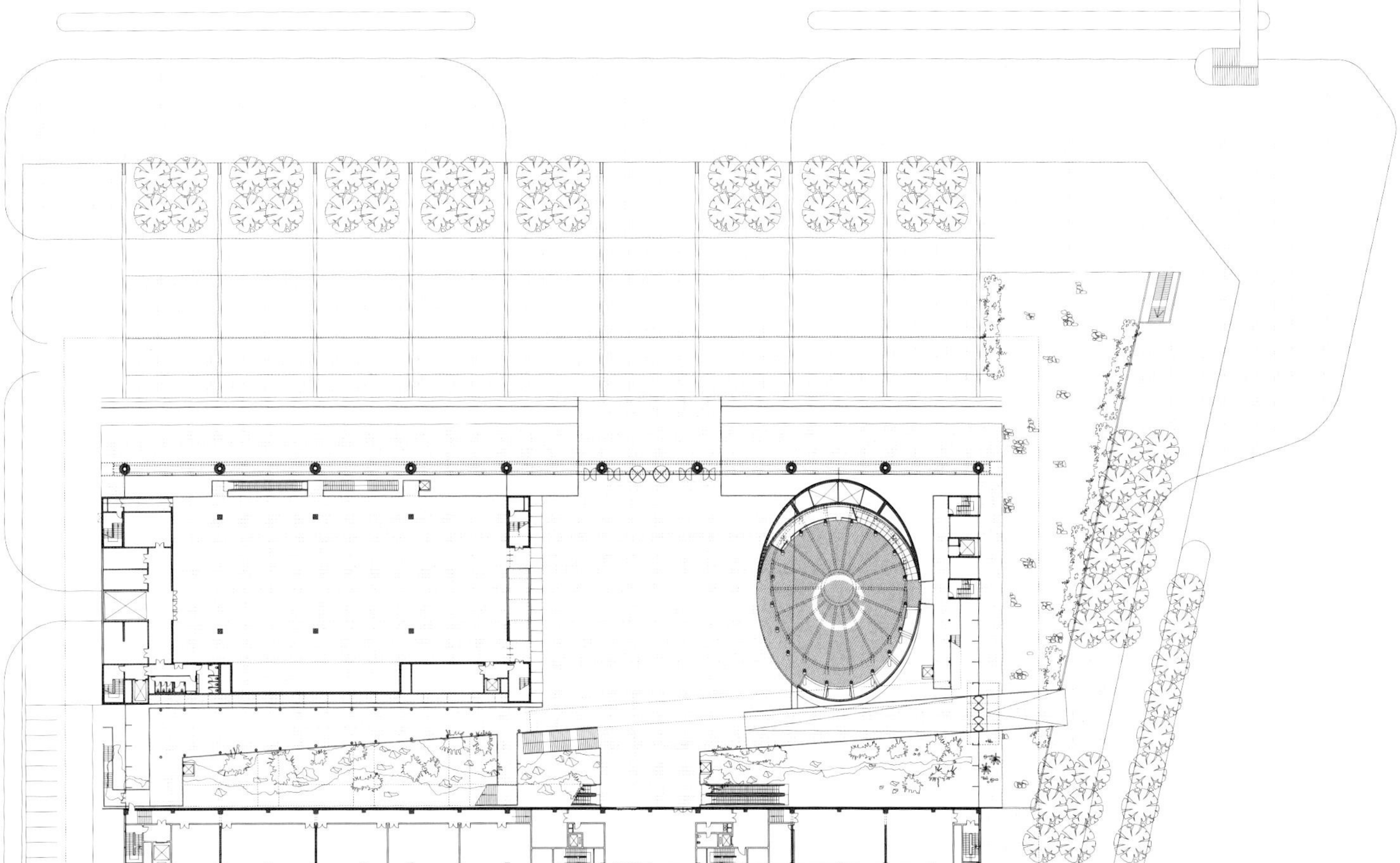

first floor

ARoS Aarhus Museum of Modern Art
아로스 현대미술관

Schmidt, Hammer & Lassen
스미트, 하머 앤 레슨

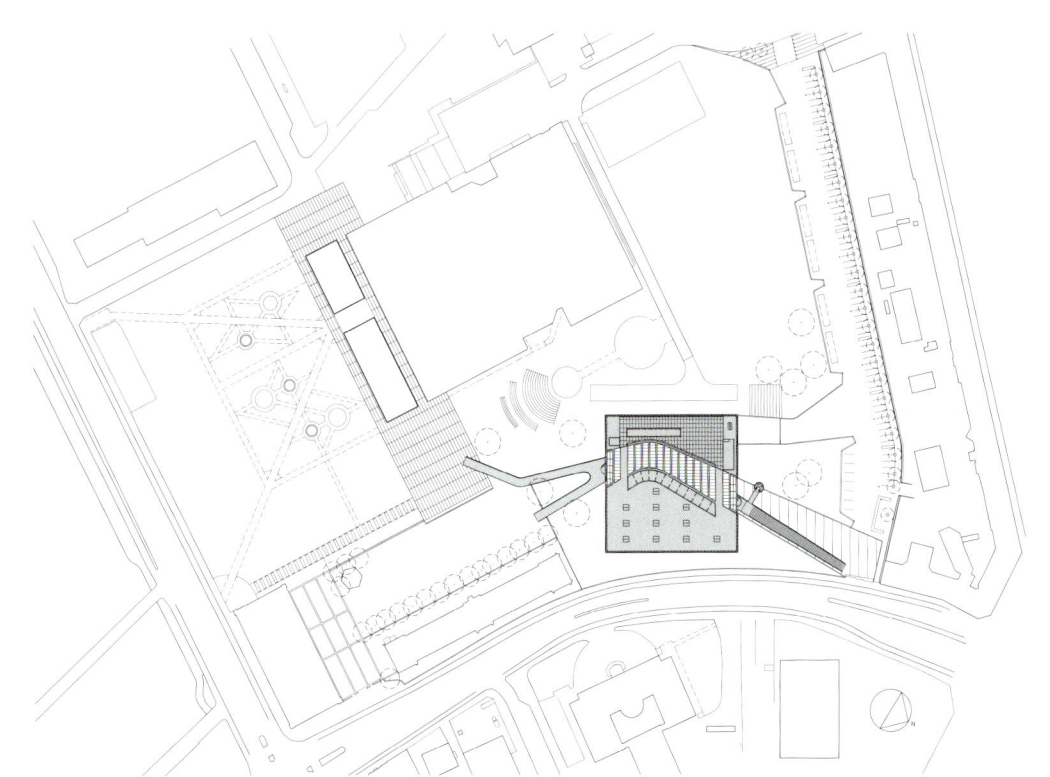

ARoS stands as a sharply defined, solid cube on red brick, measuring 52 x 52m at ground level and almost 50m high, and it is partly embedded in the sloping green land on the site. All the functions of the museum are contained in this one geometric block, which is divided by a 35m high curved section through nine of the museum's ten storeys. The curving section through the red cube of the museum forms the backbone of the museum: the expressive white inner street space where the ceiling is high and in fact the sky forms the roof. The two entrance faces of the museum on the north and south sides are linked by this museum street, which plays a dynamic overture to the many new impressions in the interior spaces in the rest of this great building. Outside the cubic volume of the museum, a ramp swings into a diagonally positioned walkway in a dynamic extension of the curving museum street out into the city in an inviting gesture gathering the movement from the city into the museum and sending it back to the city.

The high street space is the museum's main passage and point of dispersal, but it is also the division between the exhibition wing and the service wing. The curved section cuts across logically and transparently dividing the cube of the museum into a large exhibition area east of the street space and a smaller service area housing the library, offices, depots, restaurant and cafe etc. to the west.

On each side of the museum street there is an extensive foyer space : the arrival level in the museum construed as an open street and square, where various functions are expressed in the 'street furniture'. On one side of the street is the circular zone with the museum shop and cafe. On the other side are the toilet and cloakroom facilities, information desk and ticket administration area.

The foyer is the upper level in the double-height horizontal section, 'cutting' in from Vester Alle through the red cube to form a horizontal 'band of activity' between the upper and lower exhibition storeys of the building. The foyer storey is formed as an insert in visual contact with the storey below. Here it was our intention to emphasize the double height and the link with the auditorium, junior museum and the children's workshop.

In the middle of the long, gentle curve of the museum street stands the spiral stairway as a logical sculptural fixed point to bring visitors up and down to the long swing of the balconies and on to all the exhibition galleries. Two light, transparent elevator columns have been placed at the centre of the spiral stairway.

Following the spiral stairway upwards, the visitor comes to the museum's permanent exhibitions, divided over three storeys supplemented by the library, restaurant, lounge and cafe. At the very top level is a roof terrace. Downwards, the spiral stairway leads below the horizontal section in double height to more exhibition storeys : the special exhibition and the Nine Rooms of art installations.

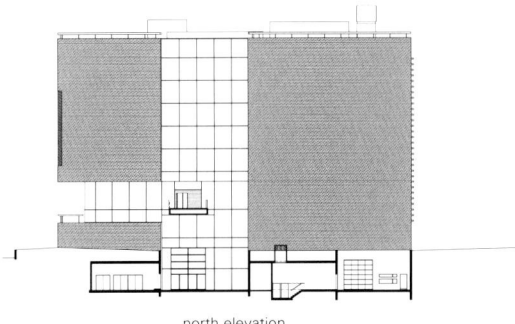
north elevation

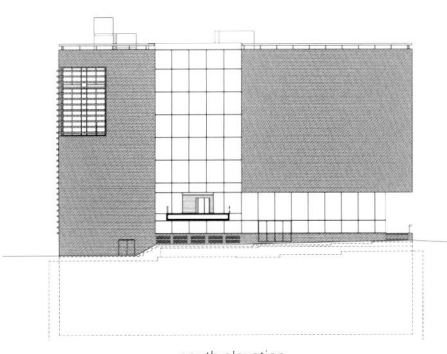
south elevation

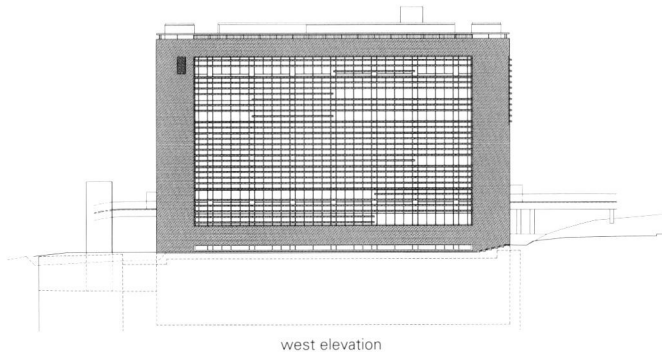
west elevation

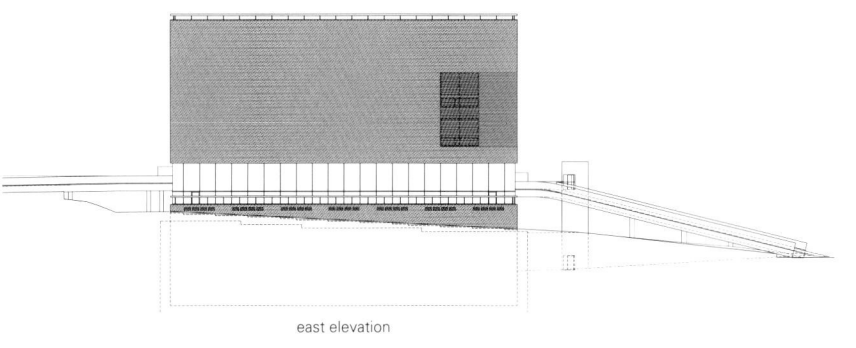
east elevation

아로스 현대미술관은 단단한 입방체형의 붉은 벽돌건물 52 x 52m, 높이 50m 로 완만하게 경사져 있는 초록의 대지에 살짝 파묻혀 있다. 총 10층의 입방체형 미술관 건물 중 9층까지는 높이 35m의 곡선형 단면에 의해 분할되어 있다. 미술관의 척추 역할을 하는 곡선형 단면은 내부의 감성적인 흰 '길'을 형성하고 천장이 높은 이 공간은 하늘을 지붕으로 삼고 있다. 남쪽과 북쪽에 있는 두 개의 출입구는 미술관 내부 길과 연결되며, 그 길은 다른 내부공간에서 보여지는 색다르고 독특한 풍경에 대한 전주곡이 된다. 입방체형 건물의 외곽에 있는 램프는 비스듬히 경사진 보도로 이어지면서 곡선형 '길'을 도시와 역동적으로 연결하는 역할을 한다. 길은 도시를 미술관으로, 미술관을 다시 도시로 이어주는 움직임을 유발하는 매혹적인 장치다.

길은 미술관의 주요 통로이자 각각의 내부 공간으로 통하는 길목인 동시에 전시구역과 부대시설 구역을 나눠주는 역할도 한다. 곡선형 단면은 미술관 블록을 가로지르면서 길의 동쪽에 있는 대규모 전시구역과 서쪽의 도서관, 사무실, 창고, 식당, 카페 등의 소규모 부대시설구역을 투명하게 분할한다.

미술관의 입구 양쪽에는 널찍한 로비가 있다. 미술관 출입구인 이곳은 '도로 시설물'로 표현되는 다양한 기능들을 갖추고 있는 열린 길과 광장이다. 길의 한쪽은 기념품점과 카페가 있는 원형구역이 있고, 그 맞은편에는 화장실과 물품 보관소, 안내 데스크, 매표소가 있다.

붉은 입방체형의 투명한 허리부분은 복층의 로비를 형성하고 있다. 로비는 베스터 알레에서 시작되어 붉은 입방체형 외관을 꿰뚫고 지나가면서 내부 전시관들 사이에 활동 공간을 형성한다. 로비층은 그 아래층과 시각적 연계를 형성한다. 대강당, 소규모 미술관, 어린이 워크숍과의 연계를 강조하는 한편 복층 구조를 부각시킨다는 것이 우리의 의도였다.

길고 완만한 곡선의 길 중앙에는 나선형 계단이 있다. 방문객들은 계단을 통해 길게 이어지는 발코니와 전시실로 출입할 수 있다. 두 개의 가볍고 투명한 엘리베이터가 나선형 계단의 한가운데에 있다. 나선형 계단을 따라 올라가면 3개 층에 걸쳐 있는 상설전시실과 도서관, 레스토랑, 라운지, 카페 등의 부대시설이 나오고, 건물 최상층에는 옥상 테라스가 있다. 나선형 계단을 따라 복층 규모의 투명 공간 아래로 내려오면 특별 전시관과 제9관까지 있는 설치미술관으로 구성된 일반전시실들이 나온다.

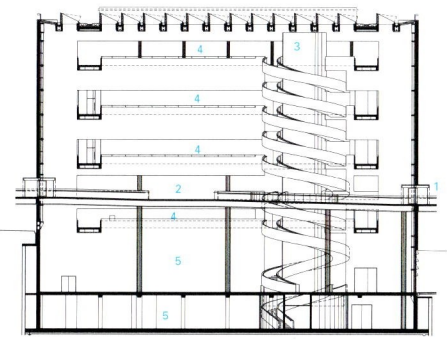

1. entry 2. street 3. stair/elevator tower 4. gallery corridor 5. exhibition

1. street 2. foyer 3. exhibition 4. changing exhibition
5. painting storage 6. gallery corridor 7. nine room

Architect : Schmidt, Hammer & Lassen
Engineer : NIRAS
Acoustic engineer : Jordan Acoustics
Landscape architect : Schmidt, Hammer & Lassen
Building consultancy service : Byggeplandata as
Client : Municipalty of Århus
Location : ARoS Allé 2, Århus, Denmark
Site area : 17,700m²
Photograph : Adam Mørk

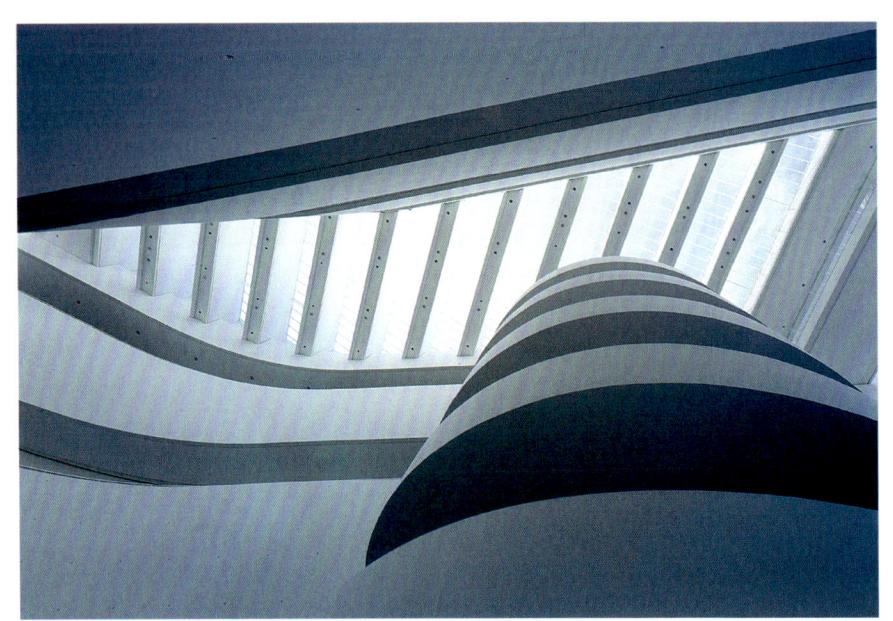

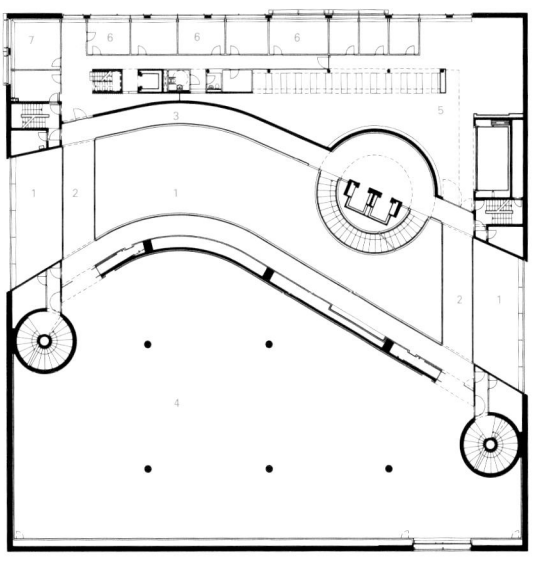

1. open   2. foot bridge   3. gallery corridor
4. exhibition   5. library   6. office   7. meeting room

level 3

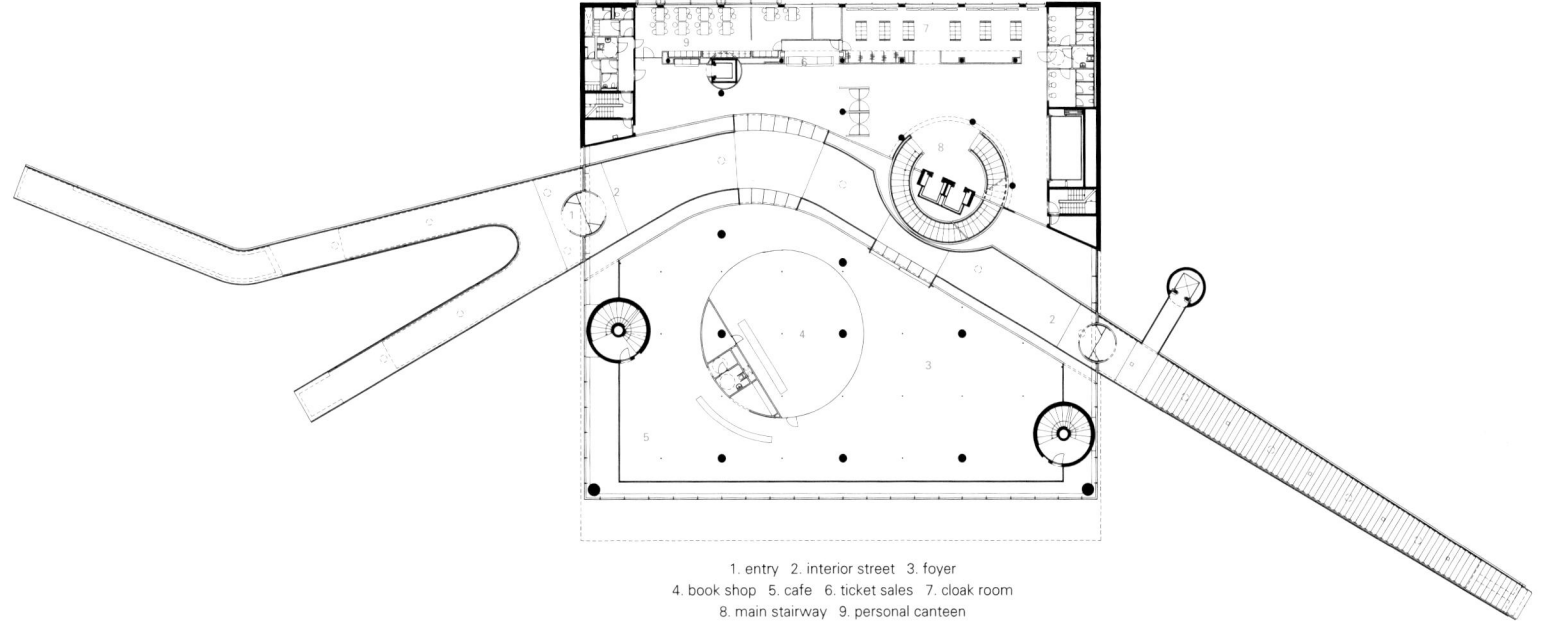

1. entry   2. interior street   3. foyer
4. book shop   5. cafe   6. ticket sales   7. cloak room
8. main stairway   9. personal canteen

level 1

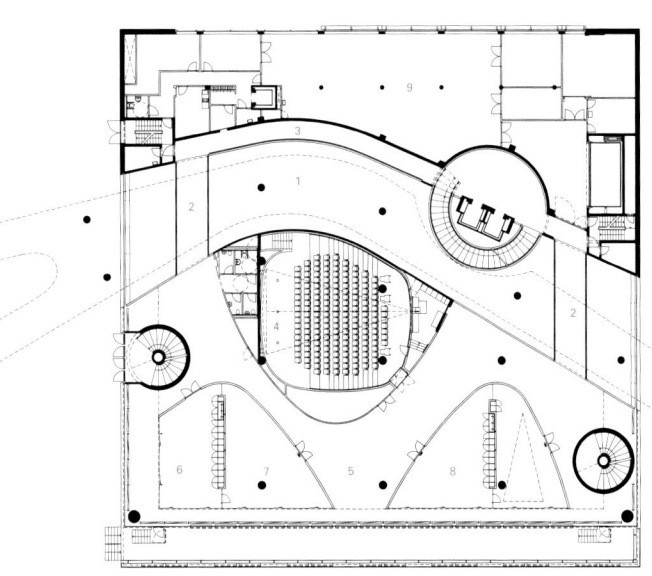

1. open   2. foot bridge   3. gallery corridor
4. auditorium   5. children's square   6. workshop   7. exhibition
8. classroom   9. conservator atelier

level 0

Palmach Museum of History

팔마치 역사 박물관

Zvi Hecker

쯔비 헥커

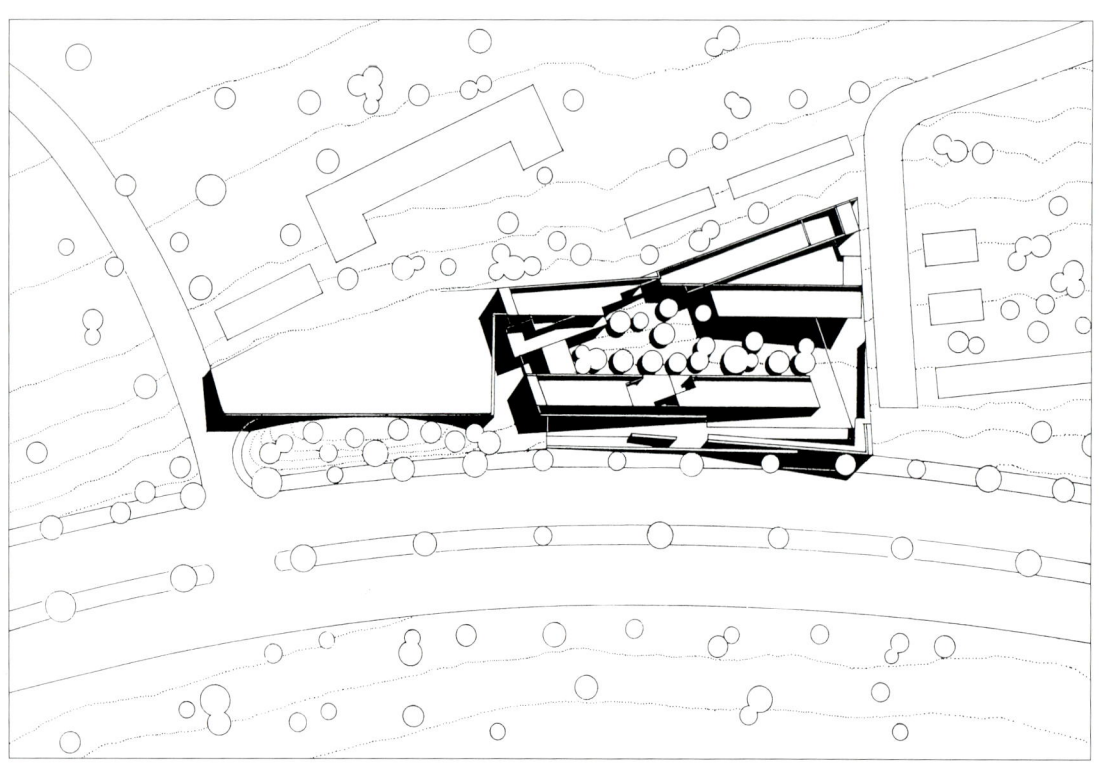

34 MUSEUM & EXHIBITION SPACE _ Palmach Museum of History

Legend Depicted in Stone

The visual representation of the Palmach [the underground resistence organization of pre-state Israel] and the enduring legend it has created is expressed by a sequence of fragmented structures embedded in the ground and partly seen from the outside. These gently wrap around a central courtyard whose existing pine trees provide its central focus. The building can be viewed as an artificial landscape comprising a series of terraces and walls ascending and mirroring the stepped nature of the terrain. The terraces reach towards the elevated regions of the site and overlook the Mediterranean sea.

As the plan of the museum derives its form from the existing site, so the elevations are conceived as extensions of the stepped terrain. The imprint of the landscape on the design is further accentuated by the use of stone discovered during the excavations for the basement of the building. Due to the decision to preserve the trees growing on the site, the museum, buried underground, is located on a narrow strip around the trees.

The ground work demanded the use of complex construction methods in order to save the growing pine trees and their surrounding soil. The ground in the Tel-Aviv region, in particular that in close proximity to the sea is largely made up of sand. It is also common to find, 'Kurkar', a soil condition which is neither sand nor sandstone but is comprised of layers of fragile sandy slates that would in a process lasting millions of years eventually transform to stone. Kurkar has never before been used as a building material. It has previously been used only as load bearing sub-layer in the construction of road surfaces and highways. The excavations around the trees demanded great care and as a result a large quantity of Kurkar was recovered. The idea of applying to the elevations found materials from the site greatly appealed to us since it encompasses the nature of the Palmach ideology and its relationship to the ground soil of Israel.

Experiments testing the application and durability of the Kurkar facade were carried out on the walls of the court-yard. Initially Kurkar layers were inserted into the wet plaster, however, the process was elaborate and time consuming. This eventually brought about the idea of applying the material more directly to the building structure and was used throughout the street elevation. The various methods and stages of experimentation became an integral part of the design of the Palmach Museum.

Beyond its powerful sculptural presence the use of the Kurkar evokes in every Israeli a vivid image so strongly related to the Palmach legend which no other building material could have achieved so directly.

Collaborator : Rafi Segal
Structural engineer : Waintraub-Naginski-Zeldin
Contractor : Solel Boneh
Client : Palmach Vaterans Association
Location : Ramat-Aviv, Tel-Aviv, Israel
Site area : 6,000m²   Bldg. area : 5,100m²
Structure : Concrete and concrete blocks
Finishing : Local sandstone, plaster,
exposed concrete plaster
Program : Exhibition space, Theater [400 seats],
Classrooms, Administration, Cafeteria etc.
Budget : 13.5 mio DM
Photograph : Michael Krüger

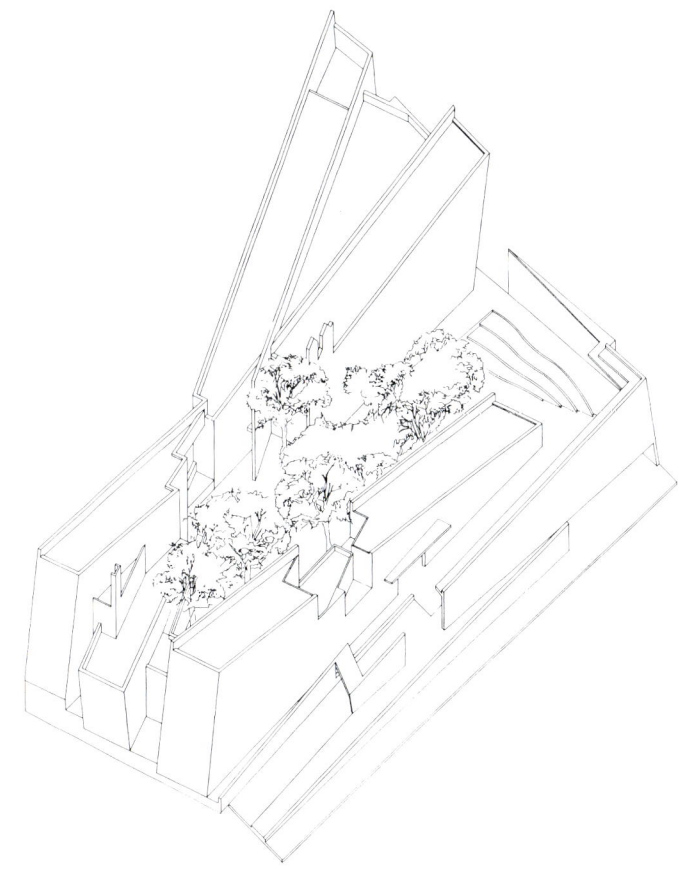

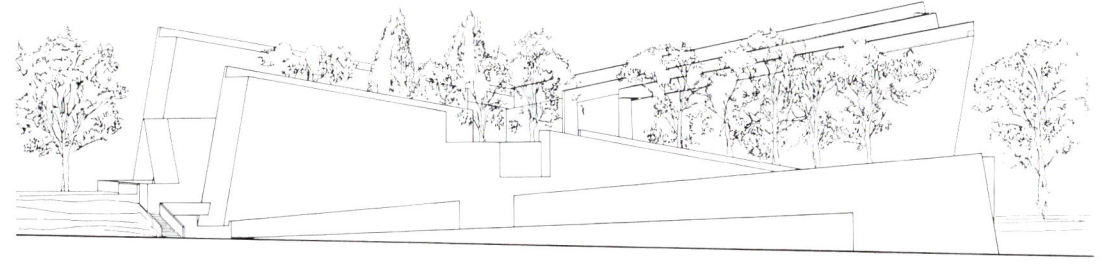

front elevation

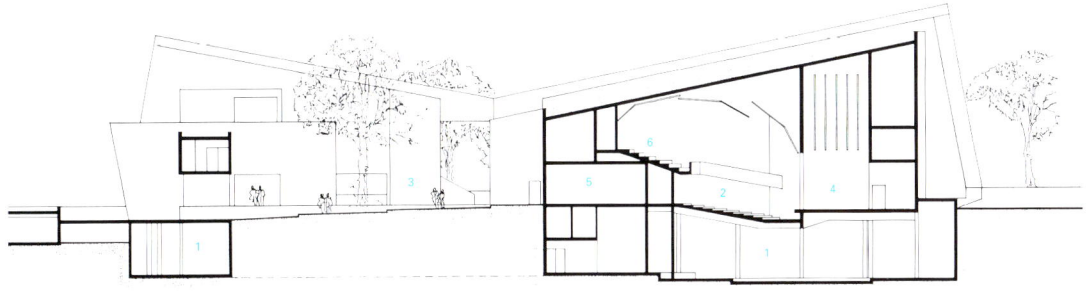

1. exhibition space   2. auditorium   3. courtyard   4. memorial space   5. lobby   6. gallery

longitudinal section

43

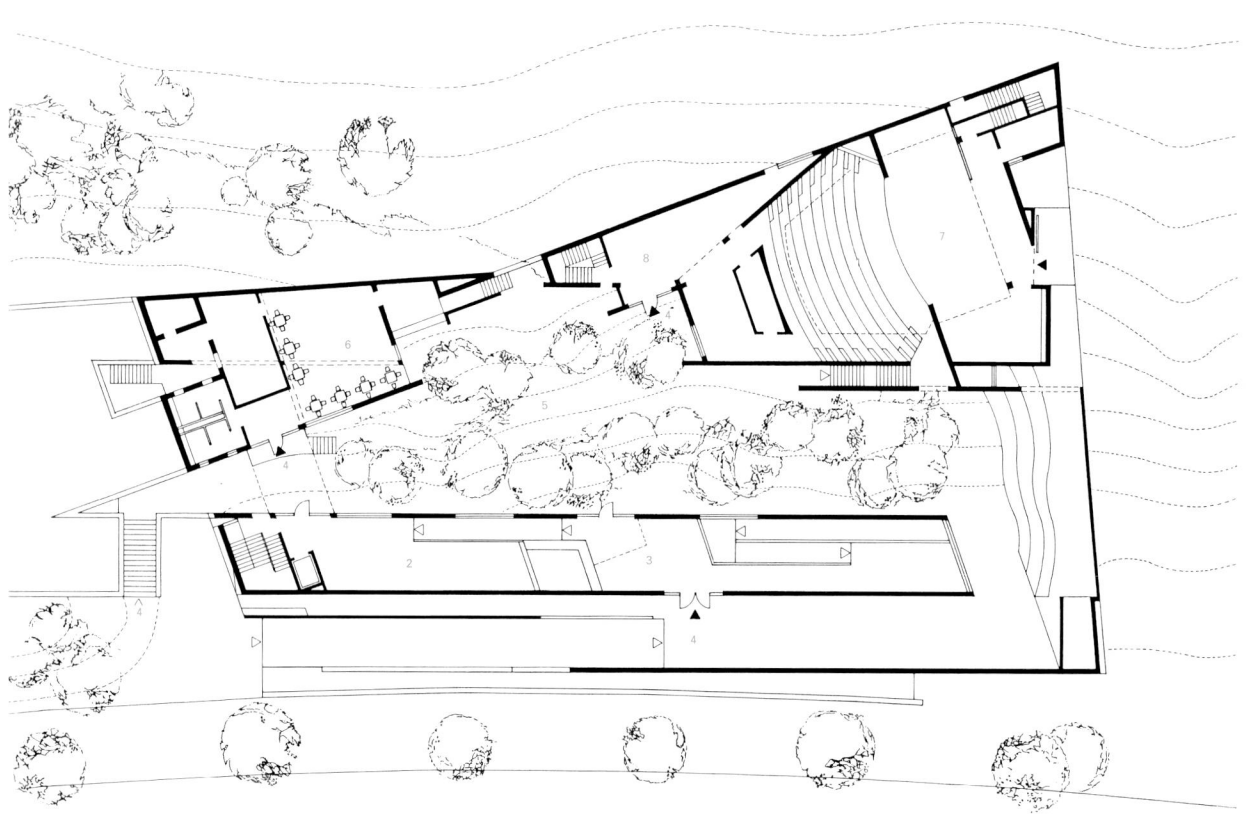

1. auditorium 2. administration 3. reception 4. entrance 5. courtyard 6. canteen 7. memorial space 8. lobby
ground floor, entrance level

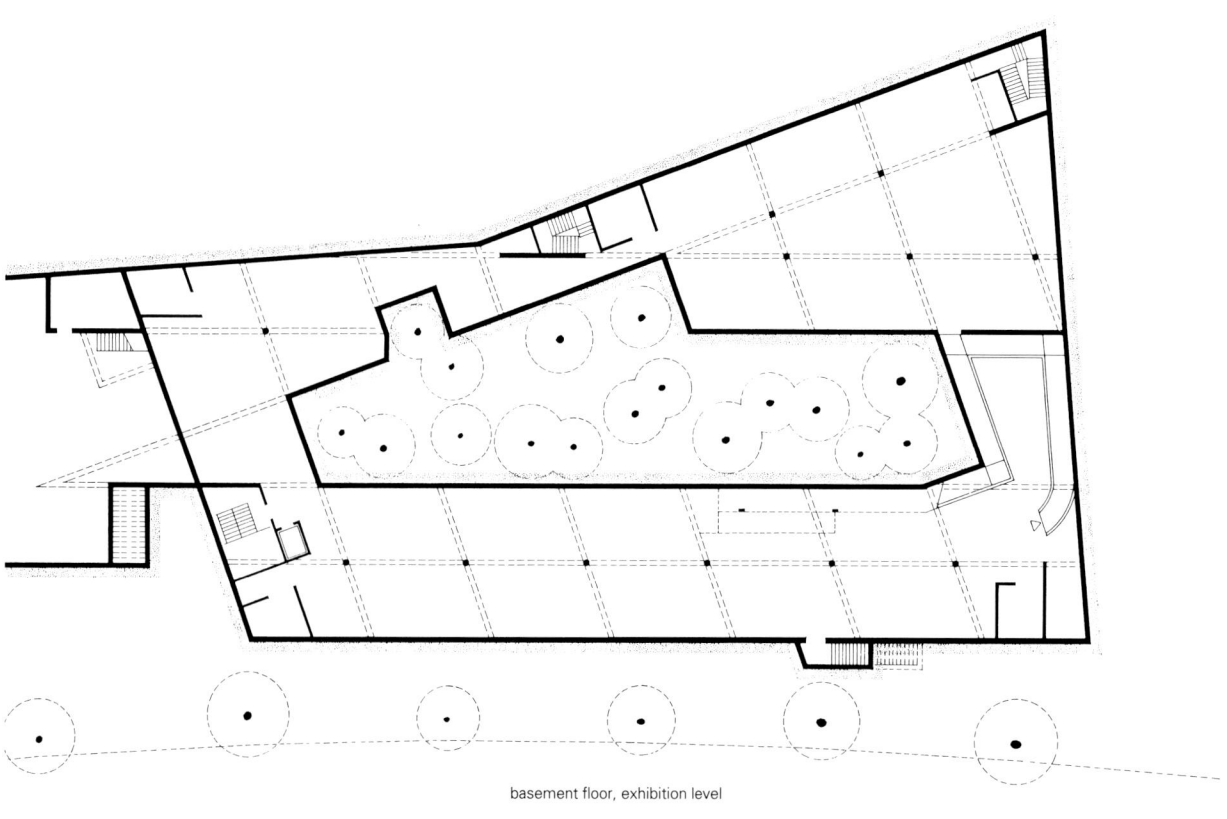

basement floor, exhibition level

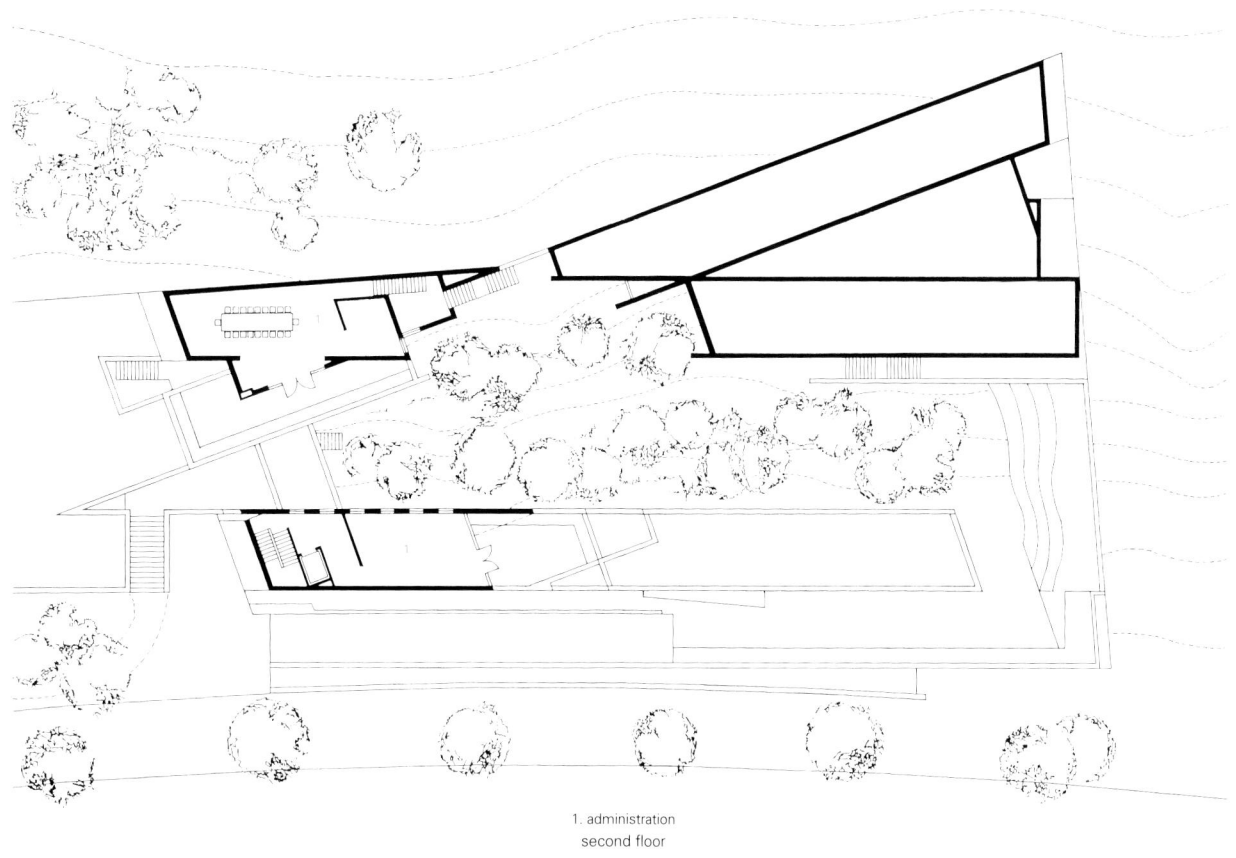
1. administration
second floor

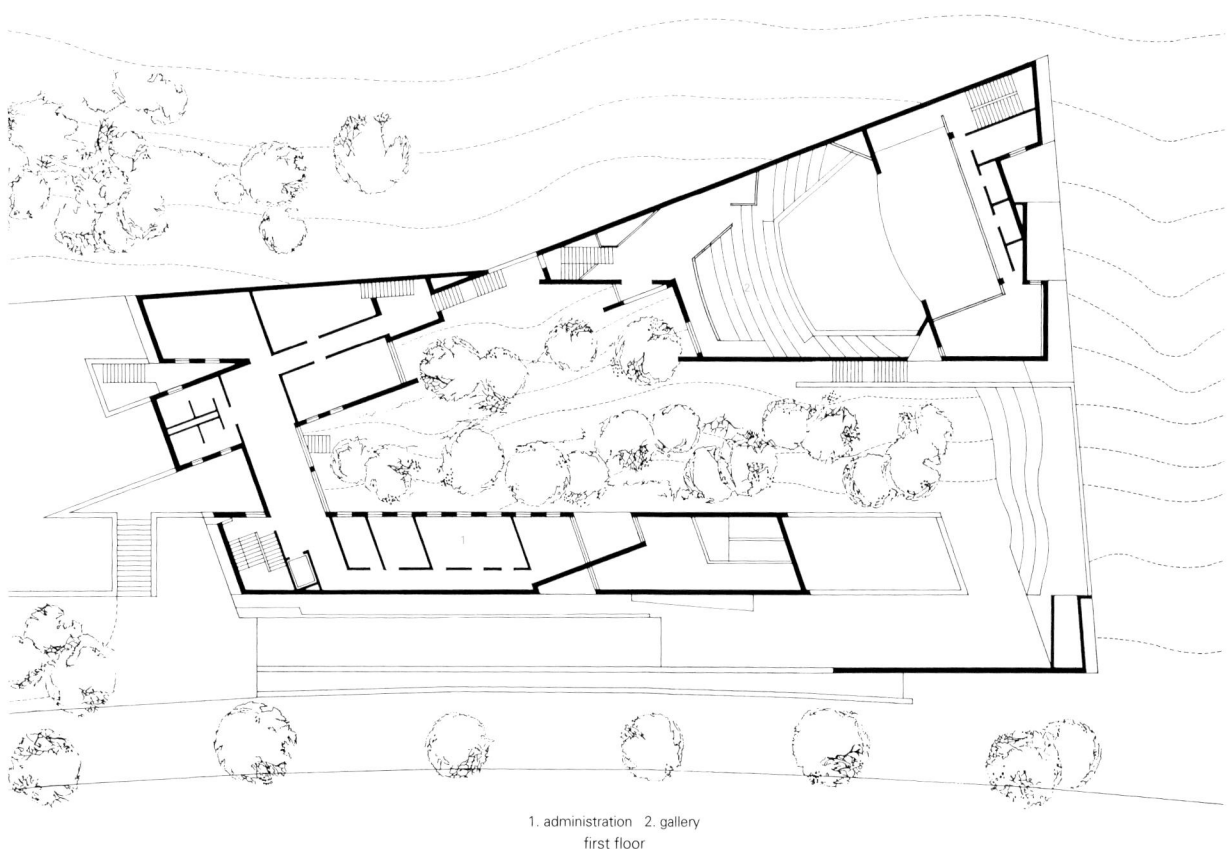
1. administration 2. gallery
first floor

돌에 새겨진 전설

국가 상립 이진 이스라엘의 지하지항조직인 팔마치와 그들에 관한 불후의 전설이, 땅 속에 묻혀 밖에서는 부분적으로만 보이는 단편적 구조물들로 우리 눈 앞에 드러났다. 이 구조물들은 기존의 소나무들이 있는 중앙 뜰을 부드럽게 감싸고 있다. 테라스와 경사진 벽으로 이루어져 계단식의 지형을 드러내 보이며 인공적 랜드스케이프를 형성한다. 테라스들은 부지의 높은 지대 쪽으로 있어 지중해를 내려다 볼 수 있도록 되어 있다.

박물관의 평면이 기존의 부지 형태에서 도출되었기 때문에 입면 또한 계단식 지형의 연장으로 이해할 수 있다. 랜드스케이프를 설계에 그대로 옮긴 흔적은 지반공사를 위해 굴착 시 발견된 돌을 사용한 데서 더욱 두드러진다. 부지에 자라고 있던 나무들을 보존하기로 결정함에 따라 땅 속에 묻혀 있던 박물관은 나무 주위의 좁고 긴 땅에 놓이게 된다.

기존의 소나무들과 주변 토양을 살리기 위해 기초공사에 복잡한 공법이 필요했다. 텔아비브 지역, 특히 바다에 인접한 땅은 주로 모래로 이루어져 있다. 모래도 사암도 아니지만 수백만 년이 흐르면 결국 돌로 변하는 약한 모래질의 슬레이트층으로 이루어진 쿠르카르 또한 흔히 구할 수 있는 재료다. 그러나 쿠르카르는 도로 표면이나 고속도로 건설시 하중을 지탱하기 위한 하부층 지반으로 사용되었을 뿐 이전까지는 건축자재로 사용된 적이 없었다. 나무 주위의 굴착 작업에는 세심한 주의가 필요했으며 그 과정에서 많은 양의 쿠르카르가 나왔다. 그 부지에서 나온 재료로 지대를 높인다는 아이디어에 우리의 관심이 집중되었다. 그것은 팔마치 이데올로기의 특성 및 이스라엘 토양과 팔마치와의 관계를 고려한 것이다.

중정에 쿠르카르를 사용함으로써 쿠르카르를 외관에 사용할 수 있는 가능성과 견고성에 대해 실험하였다. 먼저 쿠르카르 층을 젖은 벽토에 끼워 넣었다. 그러나 이것은 정교하고도 시간을 요하는 작업이었다. 이를 통해 쿠르카르를 직접 건물구조에 사용하자는 아이디어가 나왔으며, 따라서 도로지반을 높이는 데 사용되었다. 다양한 공법과 실험단계는 팔마치 역사박물관 설계에서 없어서는 안될 부분이 되었다.

주목을 끄는 조각적 형태뿐 아니라 쿠르카르를 사용함으로써 이 역사박물관은 모든 이스라엘인들에게 팔마치 전설에 관한 생생한 이미지를 불러일으키고 있다. 이는 다른 무엇으로도 얻을 수 없는 효과임에 틀림없다.

Archaeological Museum, Saint-Romain-en-Gal
생 로멩 엥 갈 고고학 박물관

Chaix & Morel
쉑스 앤 모렐

Saint - Romain - en - Gal, on the west bank of the Rhône fifty kilometres south of Lyons, was founded a little before the start of the Christian era and grew into a thriving Roman colonial town. Intersecting roads and the river running down to the Mediterranean made it a centre of trade. No less a figure than Pontius Pilate is said to have finished his days in retirement there, at Vienne, on the opposite bank.

Archaeological interest turned to the site in the 1960s and by the 80s, what with the wealth of discoveries, the idea of a permanent museum took form. Chaix & Morel won the design competition held in 1988, and the facility - some 13,900m^2 - was officially opened in 1996.

The museum stands on the riverfront just to the north of a bridge built in 1949 ; aptly, ideas of movement and crossing also underlie its design. Starting from the bridge, a wide stone stairway rises from the embankment and tapers up to a belvedere on the roof of the first building, which houses the reception area, bookshop and cafeteria, as well as temporary exhibition space and an auditorium. A propylaeum and paved forecourt a little below belvedere level provide a monumental touch and pay tribute to the Rhône, the lifeblood of the old Roman riverport colony, by opening wide views over its moving surface. A glazed walkway links this first building, whose feel is primarily concrete and glass, to the permanent exhibition building, which is mainly in steel and glass, and stands on slender pilotis to free the vestiges at ground level, rather like the makeshift sheds erected over archaeological digs. In the rear of this building a metal ramp leads down to unearthed vestiges in brick and stone around the site and between the pilotis. The fully glazed elevations of the permanent exhibition wing are protected from the harsh light of Southern France by horizontal screen - printed brise-soleil. A plane - wing overhang adds lightness to the roof suspended over the vestiges, and is echoed at the base of the glazed wall by metal duckboarding laid down for ease of maintenance. Inside the main exhibition building catwalk mezzanines have been judiciously used to display finds such as mosaics, which are best seen from a raised position, while glazed floor openings offer plunging views onto unearthed ruins directly under the building. Signage around the museum exhibits has been reduced to the minimum, in order to focus attention on exhibits and encourage a free - moving visit. Needless to say, the visit itinerary also offers superb views of the river and the built - up area of Vienne on the opposite bank : °mute reminders of what remains the same and what has changed in the landscape.

Wooden display reconstructions such as boats, wharfs and warehouse structures, along with the old ruins and the superb exhibits, remind visitors that the Romans, besides being soldiers and bureaucrats, were also designers and builders.

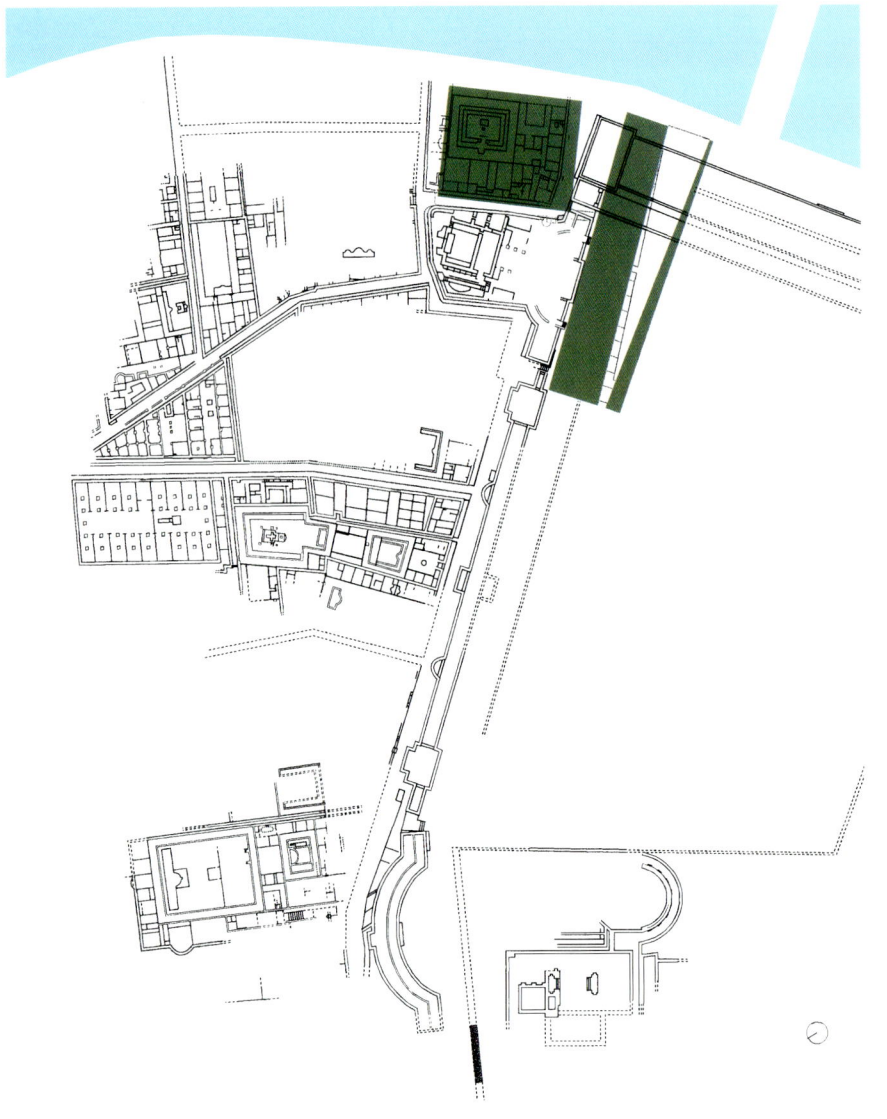

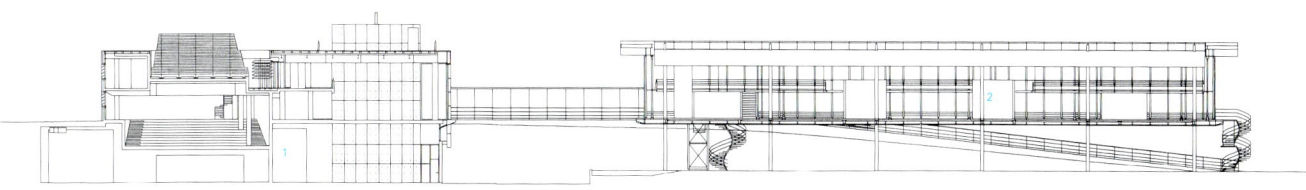

1. main building 2. permanent exhibition

cross section

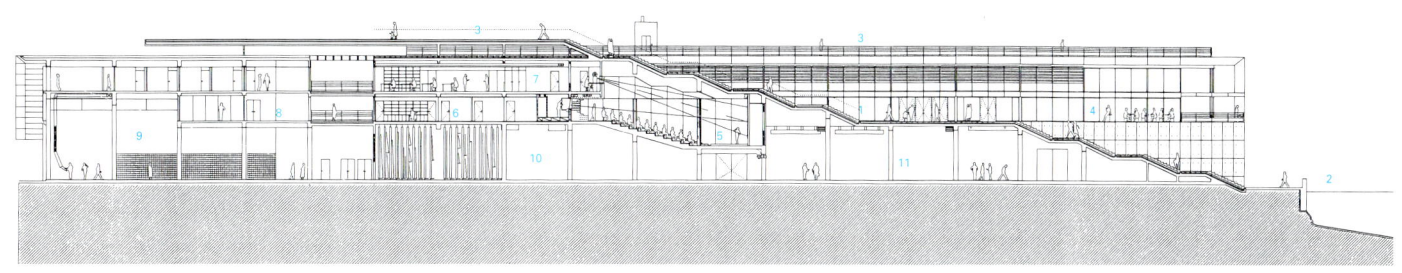

1. lobby 2. Rhone river 3. terrace 4. cafeteria & terrace 5. auditorium 6. curators' office
7. documentation center 8. archaeological research center 9. restoration workshop 10. storage

museum longitudinal section

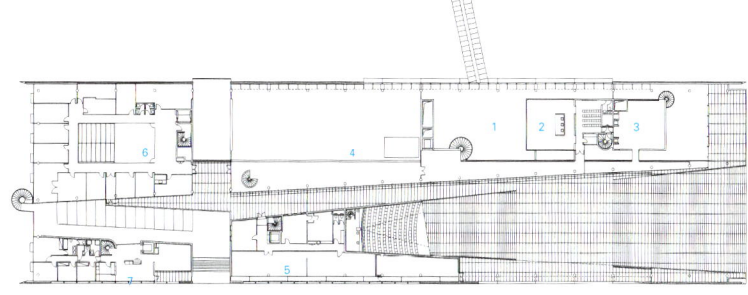

1. void   2. education service   3. cafeteria   4. temporary exhibition
5. documentation center   6. archaeological research center   7. curators' flat

mezzanine floor

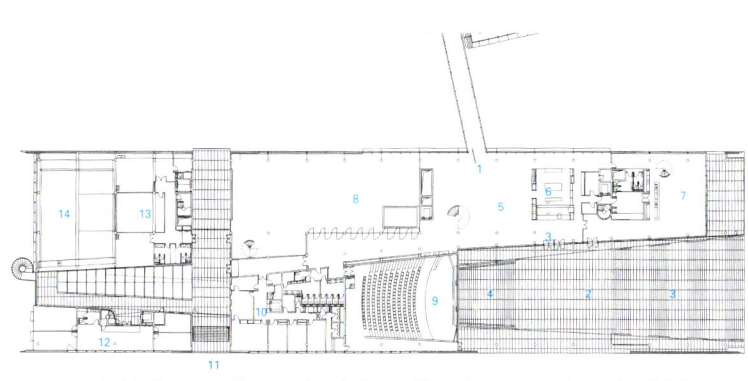

1. visitor's entrance   2. square   3. stairs from the Rhone river to terrace   4. reception   5. hall
6. bookshop & store   7. cafeteria & terrace   8. temporary exhibition   9. auditorium   10. curator's office
11. curator's entrance   12. restoration workshop   13. archaeological research center   14. void

ground floor

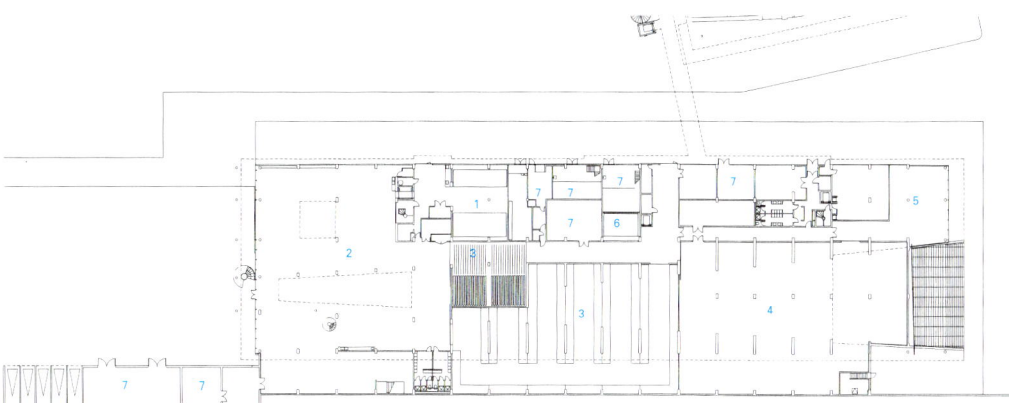

1. archaeological research center   2. restoration workshop   3. storage   4. museum storage
5. exhibition workshop studio   6. elevator   7. mechanical room

first floor below ground

MUSEUM & EXHIBITION SPACE _ Archaeological Museum, Saint-Romain-en-Gal

프랑스의 리옹 남쪽으로 50km 떨어진 론 강 서쪽 강변에 위치한 생 로멩 엥 갈은 서력 기원 직전에 건립되어 로마의 식민도시로 번영을 누렸다. 지중해로 흘러드는 론 강과 주요 교차로들 덕분에 무역 중심지로 발돋움할 수 있었던 것이다. 본디오 빌라도도 은퇴 후 맞은편 강둑에 위치한 빈에서 노년을 보냈다고 한다.

1960년대 고고학계가 이 지역에 관심을 보이기 시작했고, 이러한 관심이 80년대까지 이어졌다. 다채로운 유물과 유적이 연이어 출토되면서 상설박물관을 건립하자는 의견이 대두되었다. 쉑스 앤 모렐이 1988년에 개최된 설계경기에서 입상한 후 약 13,900m² 규모의 생 로멩 엥 갈 고고학 박물관은 1996년 정식 개관했다.

박물관은 1949년 건축된 다리의 정북쪽에 위치한 강변에 자리잡고 있다. 이동과 교차의 개념 이 설계의 밑바탕이 되었다. 다리에서 시작되는 널찍한 돌계단이 강둑으로 올라가면서 점차 좁아지고 첫 번째 건물 본관 지붕에 위치한 전망대로 이어진다. 본관 건물은 리셉션 홀, 서점과 카페테리아, 기획전시실 및 강당을 수용하고 있다. 전망대보다 약간 아래쪽에 위치한 입구와 중정은 장중한 느낌을 부여하는 동시에 유유히 흐르는 론 강 위로 드넓게 펼쳐진 전망을 제공함으로써 고대 로마 식민지의 젖줄이었던 론 강을 찬미한다. 유리가 사용된 통로는 콘크리트 및 유리를 주요 자재로 사용한 본관과 철재와 유리를 주로 사용한 별관 상설전시관을 이어준다. 지상의 유적지를 현장 그대로 보존하기 위해 날렵한 필로티 위에 있는 이 통로는 유물발굴현장에 임시로 가설된 창고를 연상케 한다. 이 건물 뒤편에 있는 금속 경사로는 필로티 사이와 대지 주변에서 발굴된 벽돌과 돌로 만든 유적으로 이어진다.

전면을 유리로 만든 상설전시관 외관은 가로방향의 스크린인쇄기법에 의해 프랑스 남부의 강렬한 태양으로부터 보호받는다. 평평한 입면으로 구성된 내물림 부분은 유적 위에 매달려 있는 지붕에 빛을 더해 주며, 손쉬운 관리를 위해 깔아놓은 금속판 길에 의해 유리벽의 기저에 반향된다. 한편 주요 전시관 내부의 좁은 메자닌층에는 주변보다 두드러진 위치에 전시해야 잘 보이는 모자이크화 같은 발굴품들이 전시돼 있으며, 유리를 이용한 바닥의 틈 사이로 건물 바로 아래에 위치한 발굴 현장을 내려다 볼 수 있게 되어 있다.

박물관 전시품들 주변의 신호체계는 최소한으로 제한, 전시품들에 주의를 집중시키는 한편 자유로운 관람을 유도하고 있다. 방문객들은 전시품을 관람하면서 론 강 및 맞은편 강변에 위치한 빈의 뛰어난 경관을 음미할 수 있다. 경치를 보며 변치 않는 것과 변화해 온 것들을 찾아볼 수 있는 일종의 침묵의 시간을 경험할 수 있는 것이다. 고대 유적지, 귀중한 전시품들과 아울러 선박, 부두, 창고시설 등과 같이 복원, 전시되고 있는 목재구조물들은 방문객들에게 고대 로마인들이 뛰어난 군인이며 관료인 동시에 탁월한 건축가였음을 다시금 실감케 한다.

Client : The General Council of the Rhone, DDE du Rhone
Location : Saint-Romain-en-Gal, France
Photograph : Christian Richters

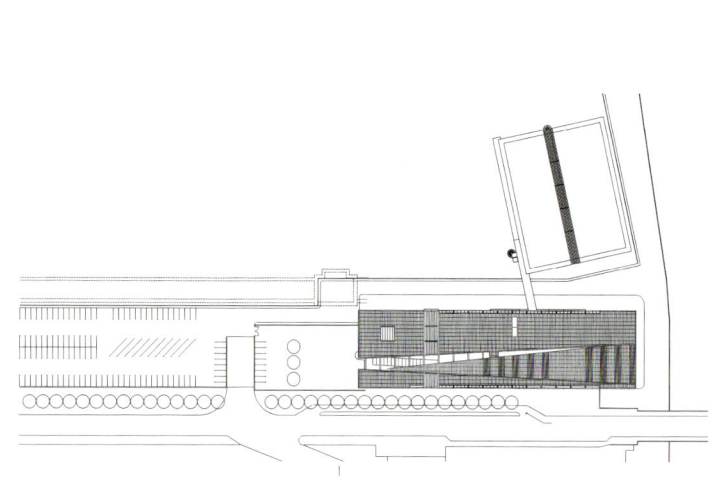

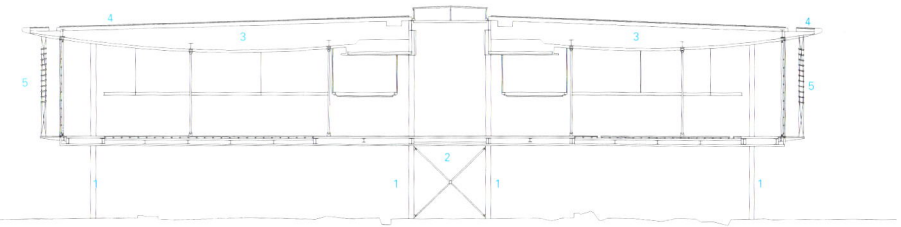

1. post on micro-pile 2. shutter 3. PRS beam 4. facade structure 5. diffuser
permanent exhibition structural scheme

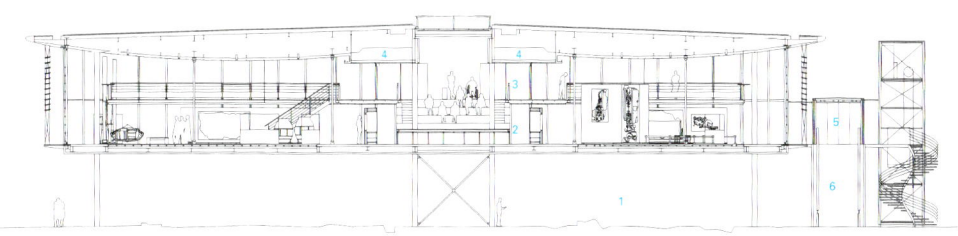

1. archaeological site 2. exhibition floor 3. mezzanine floor 4. ventilation area 5. footbridge 6. ramp to archaeological site
permanent exhibition cross section

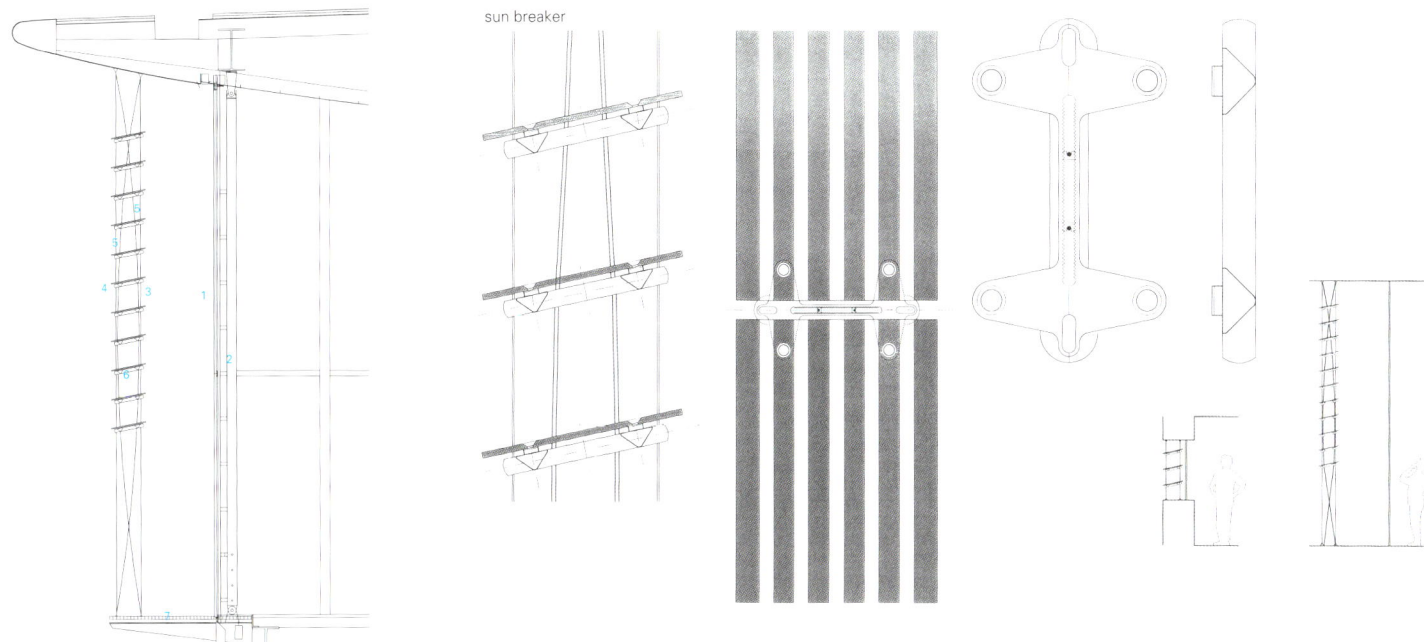

1. double glazing 2. facade structure
3. diffuser 4. silk-screen printed glazing
5. cables 6. aluminium support
7. circulation for cleaning the facade
8. conduit for convector

permanent exhibition facade sectional detail

sun breaker

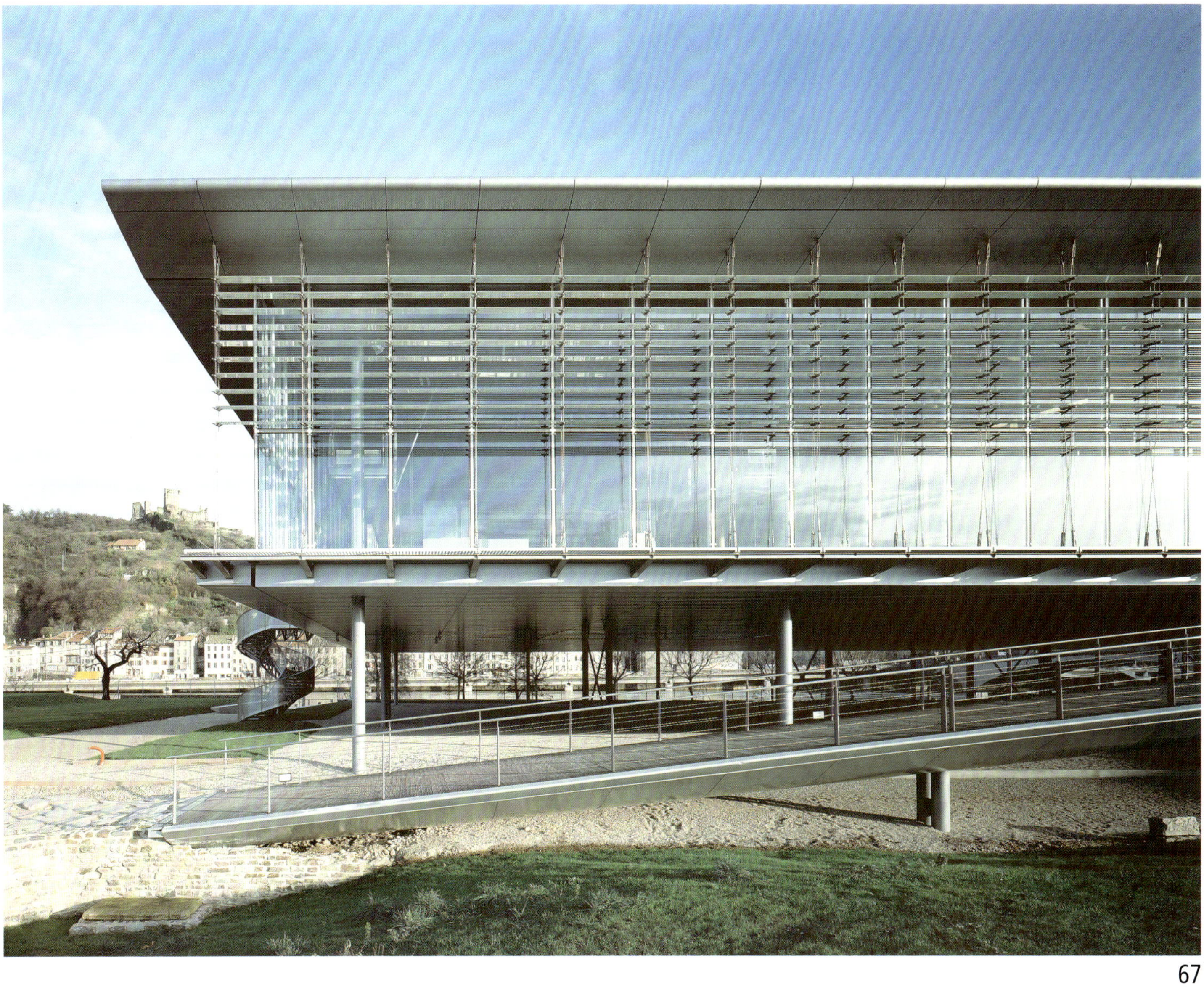

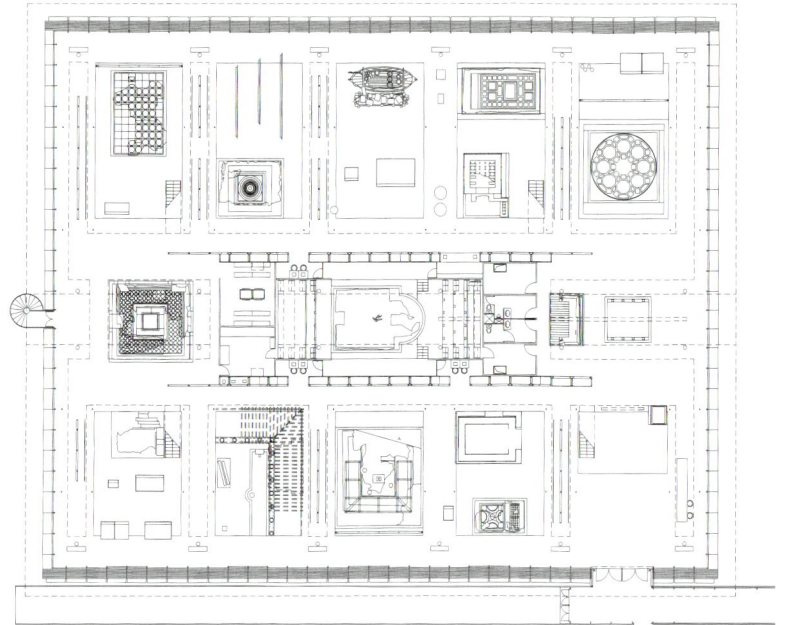
exhibition ground floor

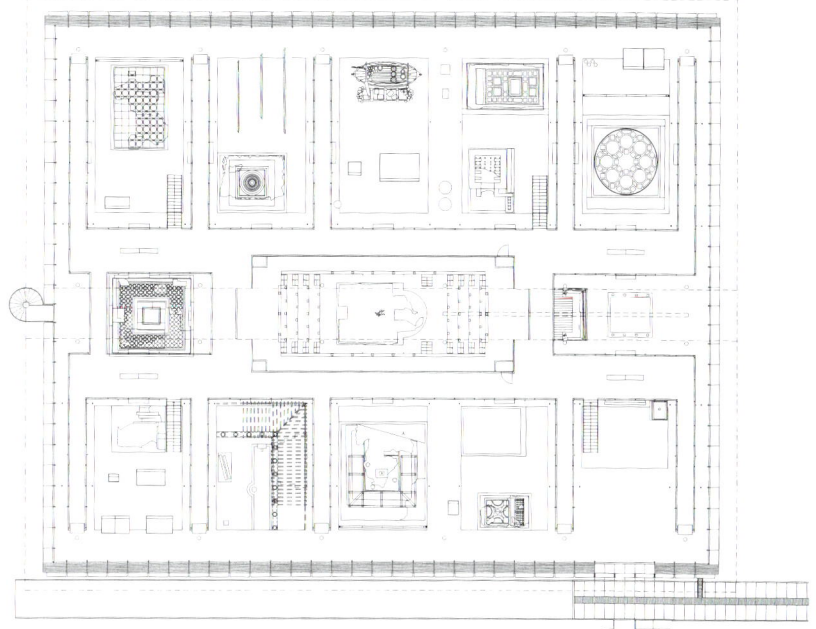

exhibition mezzanine floor

Jean Tinguely Museum
장 팅겔리 박물관

Mario Botta
마리오 보타

Works of art usually make their statements silently. Tinguely's works are an exception, for they communicate through the sounds engendered by their movements. This slight difference can, perhaps, help us understand the special demands that the works of this artist make on space. I used to think that Jean's works did not need a specially designed environment because it was inherent in their nature simply to be part of life. It is difficult to imagine a fountain divorced from its surroundings. And Jean Tinguely's works refer to emotional sources which have a close relationship with their environment and which are in constant dialogue with it.

So why build a museum? In time, the heritage of an artistic development acquires a unity which cannot be reduced to the sum of the statements of the single works. The totality of an artist's creativity opens up a perspective which is detached and different from the interpretations of the individual works of art. The assessment of the whole is dissociated from the artist's creative emotions, and often it turns out to be hard and ruthless; for, when all is said and done, it represents a confrontation with and reap-praisal of a whole historical period, of the conflicts and contradictions of a whole epoch which complement and qualify the values of the artistic statement.

In addition, the inheritance of a fortune immediately raises the question of its preservation - in fact an unfulfillable task, for it presupposes that the works of man can overcome the laws of life. The construction of a museum is a manifestation of an eternal dream, of a striving to overcome the laws of natural decline and decay to which everything is subject.

Moreover, the thought that the most intensive emotions which have accompanied and enriched our lives cannot be preserved and passed on to other people and future generations is hard to bear. Thus the legitimate wish to outlast time is intrinsic and essential to a work of art. Architecture can also be regarded as an unfulfillable task. The reasons for creating a work of architecture elude all conceivable technical, economic, functional and aesthetic motivation. The primary and ultimate reason for the existence of a work of architecture is simply to subsist as the hope of a comprehensive plan.

The adventure of the construction of the Museum Jean Tinguely in its present form and situation dates back to an idea of Fritz Gerber's. After a considerable amount of work had been done on the design of the new museum which was to be located in Frenkendorf near the home of Paul Sacher in the beautiful countryside around Basel, Gerber suggested that the building should be situated in Solitude Park on the right bank of the Rhine. The idea was immediately greeted with enthusiasm and interest by the municipal authorities, and the architects saw themselves confronted by new perspectives and challenges.

This location puts the Museum Jean Tinguely among the city's institutions, on the same level as the other establishments $^{\text{churches, schools and theaters}}$, and it represents an enrichment of the town's community life and cultural amenities.

To the architects, the construction of a museum is first and foremost the construction of a part of the city.

Its position on the right bank of the river, where the Rhine forms the boundary of a large part of the town, means that the museum is instrumental in giving a new face to a somewhat dubious urban area on the edge of the motorway. The right-angled building covers the whole east flank of Solitude Park, and its four facades each have a different spatial relationship with their surroundings.

The towering facade facing the motorway on the east of the site - the highest part of the building comprising three stories of exhibition rooms above ground level - also acts as a noise protection barrier for the park. Facing the open green area on the opposite side, the museum consists of five connected buildings, three of which adjoin a large entrance portico opening onto the park. The interior space of the museum can be subdivided by high walls which are taken up by the load-bearing structure. In this section of the building the static load-bearing system is organized according to a previously existing 5-story underground basin construction which serves to absorb the water of the Rhine.

The north elevation, which runs parallel to Grenzacherstrasse, comprises a covered area between the road and the museum which provides access to both the park and the museum. The south side of the building facing the river is a long, raised, architecturally striking building which is separated from the main volume. It represents a kind of 'promenade' over the bank of the Rhine along which visitors pass on their way to the museum and which directs their attention to the course of the river.

The exhibition areas are composed of four differently designed zones on four different levels. The first level $^{\text{2.90m above the ground floor}}$ is reached via the 'Rhine promenade' and takes the form of a gallery-like corridor opening up onto the ground floor on one side and adjoined by exhibition rooms on the other. The end of the corridor leads to a higher level $^{7.85m}$ with a series of 'classical' rooms with daylight illumination entering through side skylights and subsequently to a lower level three meters below the ground floor with rooms for the accommodation of works of art which do not require daylight.

The exhibition tour ends with the huge 'monumental sculptures' on the ground floor, in the largest of the museum's exhibition rooms $^{30\times60m}$ which can, as I have already mentioned, be divided into five rooms looking onto the park.

예술작품은 보통 조용히 그 전달하고자 하는 사상을 전한다. 그러나 팅겔리의 작품은 예외이다. 그의 작품은 움직임에서 발생하는 소리를 통해 사상을 전하고 있다. 이런 근소한 차이를 알고 나면 이 예술가가 공간 속에서 만들어 내는 특별한 주장을 이해하는데 도움이 될 수 있을 것이다. 나는 장의 작품을 위해서 특별히 디자인된 환경이 필요하지 않을 것이라고 생각한다. 본질적으로 삶의 일부가 되는 그의 작품의 특성 때문이다. 주변환경과 분리된 분수를 상상하기는 힘들다. 장 팅겔리의 작품은 주의 환경과 긴밀한 연관을 갖고 있으면서 끊임없이 대화를 지속하는 감정의 원천을 다룬다. 그렇다면 왜 박물관을 짓는가? 일정한 시기가 되면, 예술적 발전의 유산은 개개 작품의 의미를 모두 담아낼 수 있는 총체를 획득하게 될 것이다. 이렇게 획득된 예술가의 창조성의 총체는 개개 예술작품에 대한 해석과는 다른 관점을 제시한다. 작품을 전체적으로 평가한다는 것은 예술가의 창조적인 감각과는 별개이다. 송종 그러한 행위는 어렵고도 잔인한 행위로 판명된다. 왜냐하면, 모든 평가는 갈등과 모순으로 가득하지만 그 작품의 예술적 진술의 가치를 규정하고 보완해 줄 수 있는 역사의 전 시대를 거쳐 어떻게 그것에 대응했는지, 그 속에서 어떻게 재평가되는지 나타나기 때문이다. 게다가 유산을 물려받으면 즉시 그 보존에 대한 의문이 제기된다. 그러나 인간의 작품은 삶의 법칙을 극복할 수 있다는 것을 전제로 하고 있으므로 실현될 수 없는 작업이다. 박물관 건설은 영원한 꿈의 표현이며, 만물을 지배하고 있는 쇠퇴와 부패의 자연법칙을 극복하고자 하는 노력의 표현이다.

더욱이 우리 삶과 함께하며 삶을 풍요롭게 했던 가장 격렬한 감정도 보존될 수도 타인이나 후세에게 물려줄 수도 없다는 생각은 견디기 힘든 것이다. 따라서 시간을 초월하려는 이 당연한 소원은 예술작품에 있어 본질적이고 모든 기술적, 경제적, 기능적, 어학적 동기에서 벗어난다. 건축작품이 존재하는 주요하고 궁극적인 이유는 종합적인 계획의 희망으로서 존립하는 것이다.

현재의 형태와 상황에서 이루어진 장 팅겔리 박물관 건립의 모험은 프리쯔 거버의 사상에서 비롯되었다. 바젤 주변의 아름다운 시골 마을인 파울 자허 생가 근처인 프렝켄도르프에 세워질 예정이던 새 박물관의 디자인이 상당히 진척되었을 때, 거버는 그 건물이 라인 강 오른쪽 둑의 솔리투드 공원에 지어져야 한다고 제안하였다. 이 제안은 곧 시관계자들의 열렬한 지지와 관심을 불러일으켰고, 다른 건축가들은 새로운 관점과 도전에 직면하게 된 것이다.

장 팅겔리 박물관은 교회, 학교, 극장 등과 같은 다른 공공시설물들과 똑같은 높이로 지어졌다. 이것은 그 지역사회의 삶과 문화시설의 풍요를 상징하고 있다.
건축가들에게 있어서 박물관은 우선적으로 건설되어야 할 도시의 일부이다.
이 박물관이 마을 중심지의 경계 형태가 되는 라인 강 오른쪽 둑 위에 자리한 것은 고속도로변에서 다소 모호한 이 도심지의 모습에 신선함을 주는 역할을 하고 있음을 의미한다. 이 직각형태의 건물은 솔리투드 공원의 동쪽 측면 전체를 덮고 있고, 사면의 파사드는 그 주위 환경과 각각 다른 공간적 관계를 형성하고 있다.

대지의 동쪽에 있는 고속도로와 면한 높이 치솟은 파사드는 지상 3층의 전시실을 갖춘 이 건물의 최고층으로써, 도시로부터 공원으로 오는 소음을 막아주는 장벽 역할을 한다. 그 반대편에는 녹지가 있으며, 다섯 동의 연결된 건물로 이루어져 있다. 그 중에 세 동은 공원으로 개방되어 있는 거대한 출입구 포티코에 인접해 있다. 박물관의 내부공간은 하중버팀구조로 된 높은 벽으로 다시 나누어져 있다. 건물의 이 부분은 라인 강물 흡수를 목적으로 지어셨던 기존의 지하 5층 규모의 분지 시공을 따라 정적 하중버팀체계로 구성하였다.

그렌짜허슈트라쎄와 평행한 북측입면에는 도로와 박물관 사이에 공원과 박물관 모두에 진입할 수 있는 덮인 지역이 있다. 강을 마주하고 있는 남쪽 면은 건물 본관과 분리되어 길고 우뚝 선 건축적으로 두드러지는 건물이다. 이 건물은 라인 강 둑 위를 따라 생긴 일종의 '산책로'로 나타나며, 방문객들이 그 길을 따라 박물관으로 통하도록 하고, 방문객의 시선을 강쪽으로 유도하기도 한다.

전시 공간은 네 개의 각 다른 높이에 각각 다르게 디자인되었다.
지상 2.9m의 1층은 '라인 산책길'을 통해 연결되며, 갤러리 같은 복도 형태도 한편은 1층에 다른 한편은 전시실에 인접해 있다. 회랑의 끝은 그 위층으로 $^{2.9m}$ 이어져서 측면 채광창을 통해 들어오는 일광 채광을 이용하는 '고전' 전시실들이 있고, 이에 연결되어 1층보다 3m 정도 낮은 층에는 일광을 필요로 하지 않는 작품들을 전시할 수 있는 예술 작품 수용실들이 있다.

관람은 1층에 위치한 가장 큰 전시실$^{30 \times 13m}$에 있는 거대한 기념상들을 관람하는 것으로 끝난다. 이 전시물은 이미 언급했던 대로 공원을 내다보는 5개의 실로 나눌 수 있다.

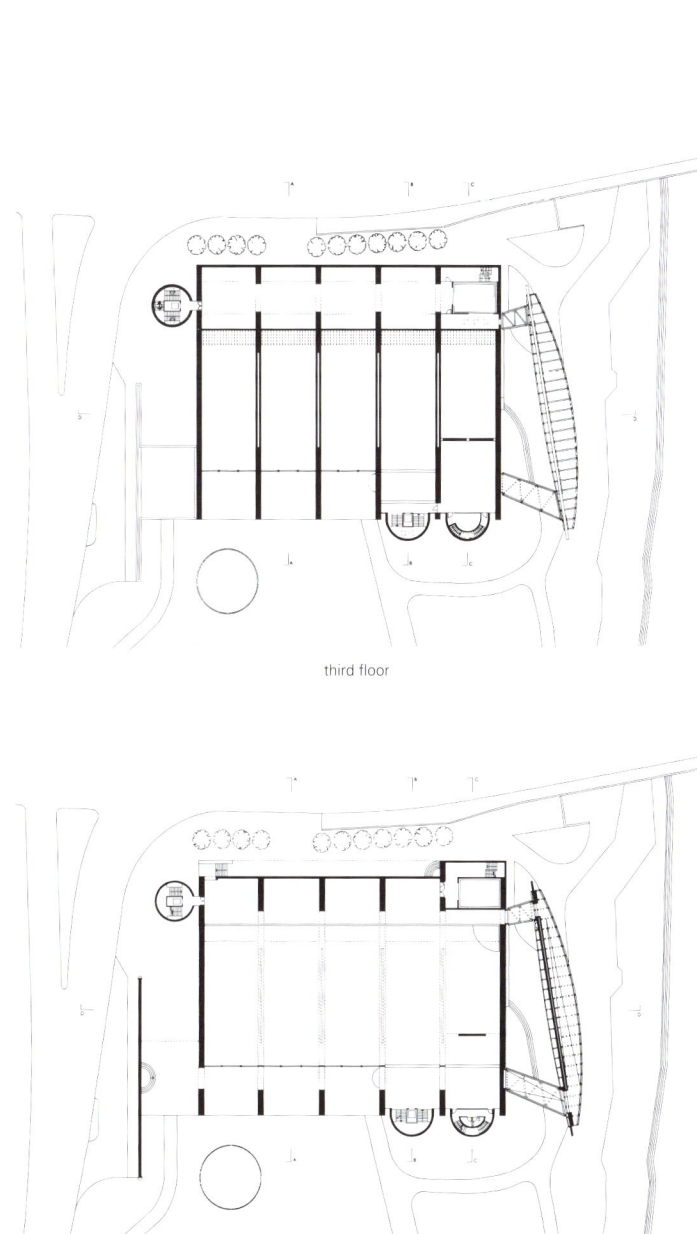

third floor

second floor

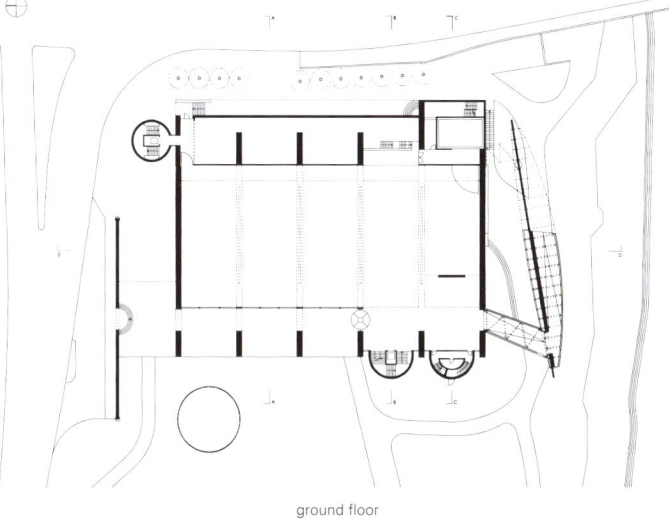

ground floor

Builder : F. Hoffmann-La Roche AG Basel
General contractor
: Georg Steiner, GSG Baucontrol AG Basel
Museum administration
: Margrit Hahnloser, Pontus Hulten
Exhibition space : 2,866m^2
Total area : 6,057m^2
External dimensions : Length/61.2m
Width/48.7m
eight/12.6, 15.4m
Floor of exhibition area
: Parquet-larch, glazed grey
Floor of coffee shop
: Parquet-oak, glazed grey
Floor of stairs
: Granite-nero angola, water-jetted
Paving of surrounding area
: Granite-Portugal
Interior walls
: Plaster and plasterboard$^{painted\ white}$,
Stucco lucido veneziano^{black}
Exterior walls
: Sandstone-Rosé de Champenay^{Alsace}
Ceilings : Acoustic panelling
/Wood, white, lignoform type,
MDF, topakustic type
Roofing of halls
: Flexible vaulted metal sheet roofing,
corrugated zinc titanium sheeting
Photograph : Pino Musi

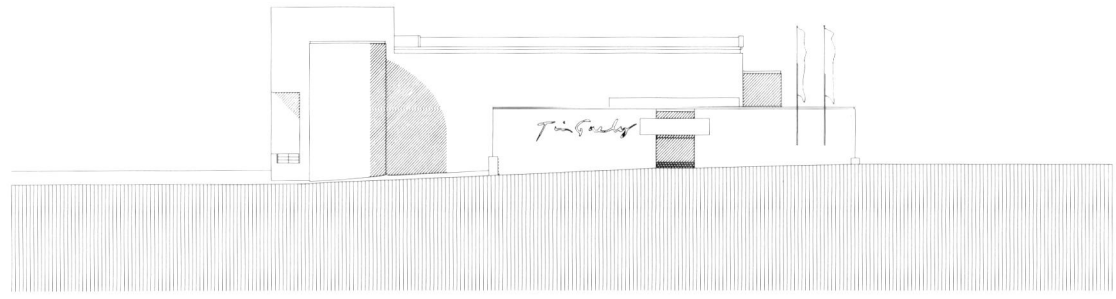

north facade

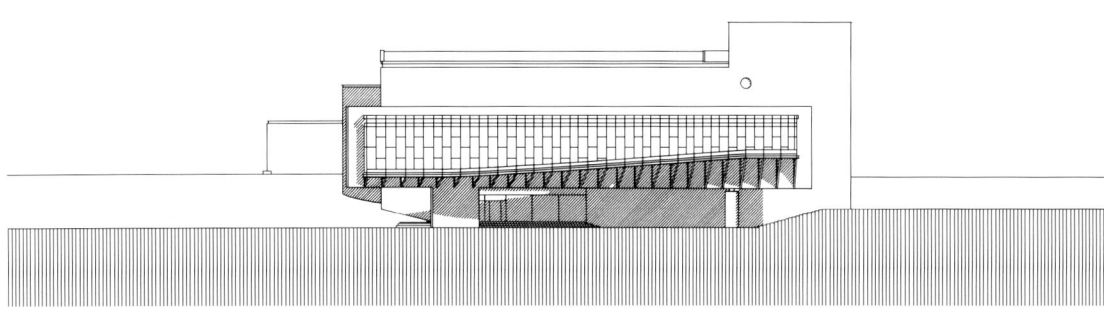

south facade

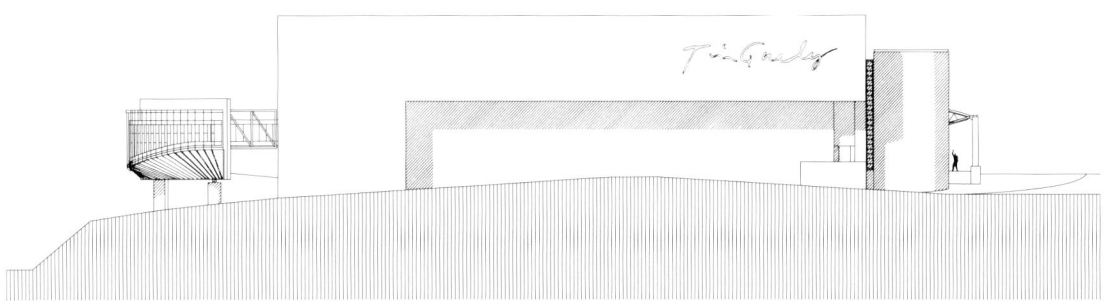

east facade

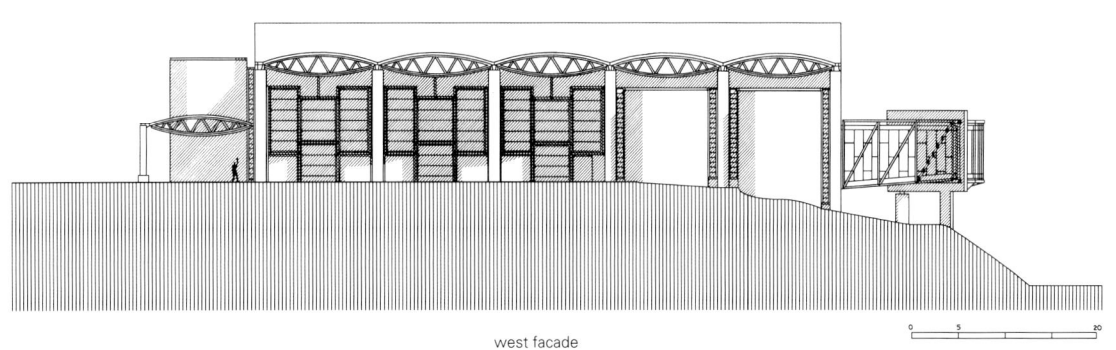

west facade

cross section　　　　　　　　　　　　　　　cross section

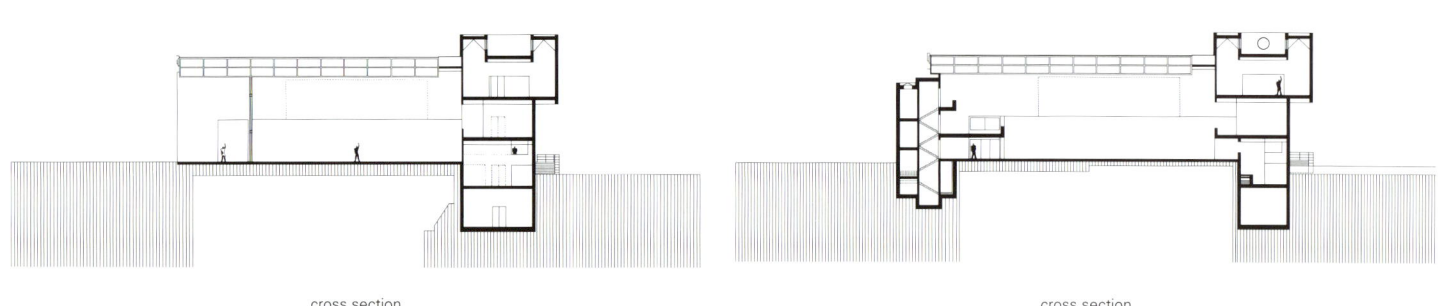

longitudinal section

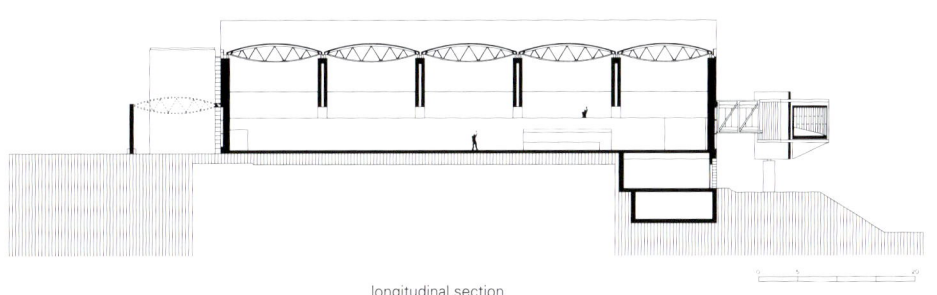

81

MuseumsQuartier Wien
빈 뮤지엄콰르티에 2제

Ortner & Ortner Architecture
오트너 앤 오트너 아키텍춰

The MuseumsQuartier is located in a revitalized grouping of Baroque royal stables in the center of Vienna. Together with the Hofburg, the Neue Burg, the Kunsthistorisches Museum, and the Naturhistorisches Museum, the area represents a historically unique force field of Austrian identity. As the largest cultural complex in the history of the Republic of Austria, the MuseumsQuartier with its numerous arches and passageways, the newly designed Forecourt, and the relaxation zones in the interior establishes a connecting axis that brings the city's individual cultural areas even closer to each other.

The defining motif for the MuseumsQuartier - both externally and internally - is the reciprocal synthesis of historic and contemporary architecture. With the design for the new MuseumsQuartier, Laurids Ortner has ingeniously connected urban energy fields to create a symphonic alliance of old and new, art and recreation, artists and audience. Architect Manfred Wehdorn is responsible for the renovation of buildings classified as historical monuments.

빈 콰르티에뮤지엄은 비엔나 시내에 위치하며 개조된 바로크식 왕국 건물내에 위치한다. 구 왕궁, 신 왕궁, 미술사 박물관, 자연사 박물관 등이 함께 어우러진 역사적으로 오스트리아를 대표하는 장소이다. 오스트리아 공화국 역사상 최대 문화 단지인 빈 콰르티에뮤지엄은 수많은 아치와 통로들이 있으며, 새로 설계된 앞뜰과 내부의 휴게 구역은 비엔나의 각 문화 영역들을 서로 긴밀하게 연결시키는 축을 형성한다.

외·내부 전체에서 빈 콰르티에뮤지엄의 모티브는 역사적 건축과 현대 건축을 상호 복합적으로 조성하는 것이다. 새로운 빈 콰르티에뮤지엄을 설계한 라우이쯔 오트너는 옛 것과 새 것, 예술과 재현, 그리고 예술가와 관객간의 조화로운 관계를 형성하기 위해 도시의 에너지 장들을 정교하게 연결했다. 건축가 만프레드 베도른이 역사적인 기념물로 평가되는 건물의 개조를 책임진다.

Architect : New buildings - Arch. Prof. Dipl. Ing. Laurids Ortner, Arch. Prof. Mag.art. Manfred Ortner
Old buildings - Arch. Univ. Prof. Dipl.Ing. Dr. Manfred Wehdorn
Building Contractor : Art - Holzmann BaugesmbH
Owner : Museums Quartier Errichtungs - und, Betriebsgesellschaft mbH,
Dr. Wolfgang Waldner, CEO
Photograph : Rupert Steiner

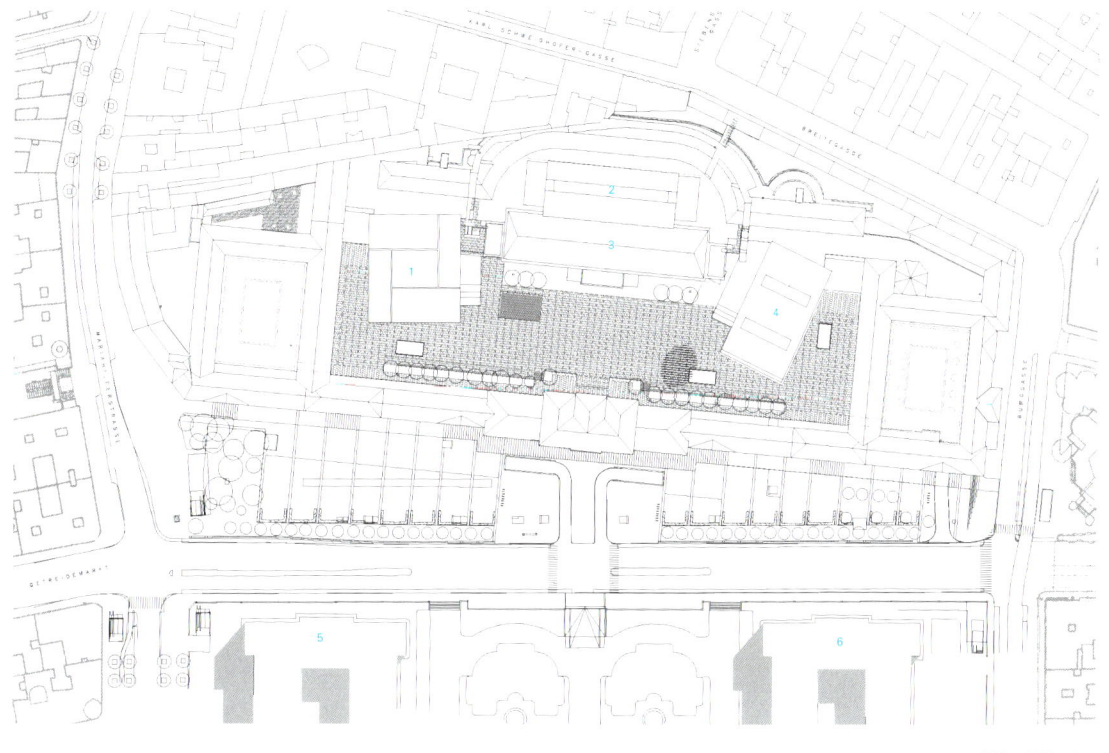

1. Leopold Museum 2. Kunsthalle Wien 3. Veranstaltung 4. Museum Moderner Kunst Stiftung Ludwig Wien
5. Kunsthistorisches Museum 6. Naturhistorisches Museum

Leopold Museum
레오폴드 미술관

As a compact cuboid measuring 40 x 46 x 24 meters and clad entirely in white limestone, the building is perfectly synchronized with its surroundings. A solid stone shell with a delicate, preciously treated surface is broken up by modular openings in the facade and roof.

A 10-meter wide outdoor stairway leads from the entrance level to 3.4m above the courtyard level. Inside, a 250m² glass-covered atrium forms the central axis for the exhibition halls situated around it. The atrium is 19m high and links the three upper levels. A lower atrium connected to the upper atrium through an opening in the ceiling is 9m high and connects the two lower exhibition levels.

From the entrance level, the main stairway on the long side leads over the mezzanine with the shop and cafe the two upper levels and the two lower levels. Another subterranean level is used for storage and utilities. Two representative, two-story-high rooms are directly adjacent to the atrium.

Childcare and educational facilities are situated in the bridge to the old building. The auditorium with its separate outside entrance is under the outdoor staircase.

The administrative offices are in the adjacent old building wing, which can be reached over the bridge. Delivery zones and workshops are in the Oval Wing and are connected to the new building through an underground tunnel.

On each floor, four large exhibition halls are positioned following a modular layout and are separated from each other by a common access core. Each hall is subdivided with partition walls depending on the respective exhibition concept. Daylight enters the four upper levels from the side or from above through equally sized window openings. In addition, three halls are illuminated through elongated windows along the upper edges of the walls. An artificial lighting system provides for even lighting of the exhibition walls. The lowest level of the museum, which has no daylight illumination, is used for special exhibition purposes. The halls are equipped with a special light ceiling.

Materials Used: White limestone ^{Vratsa} for the facade, roof, walls, ceilings, and floors in all visitor areas and the access cores; oak parquet flooring in the exhibition halls; patinated brass for all visible metal parts.

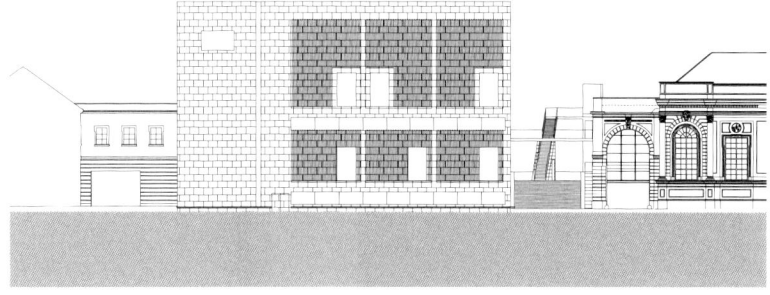

northeast elevation

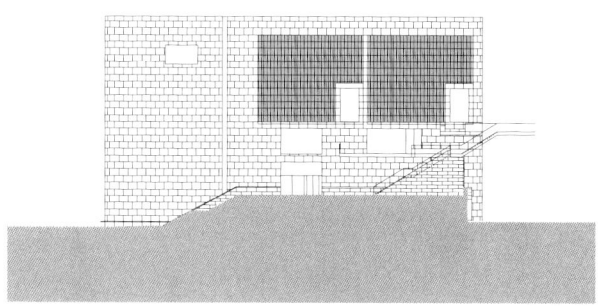

northwest elevation

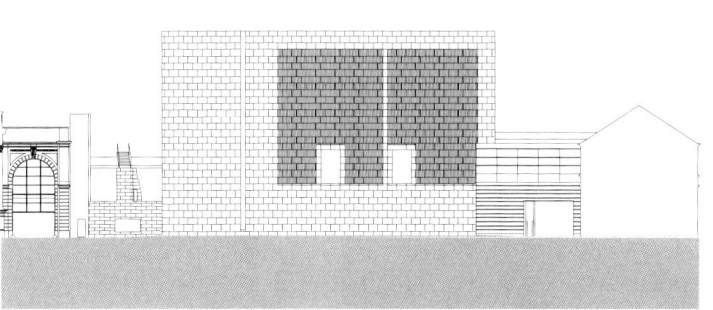

southwest elevation

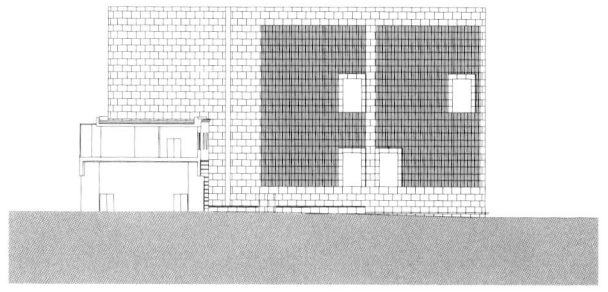

southeast elevation

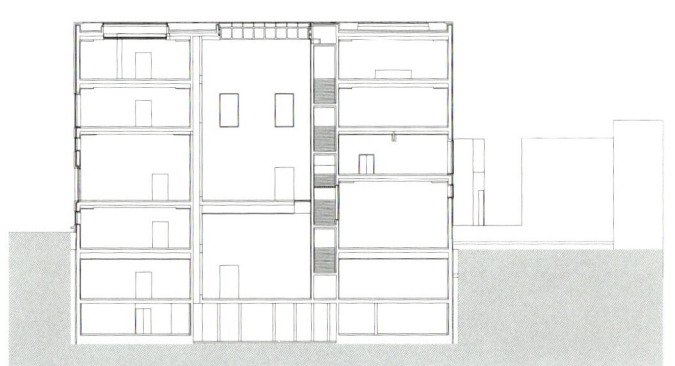
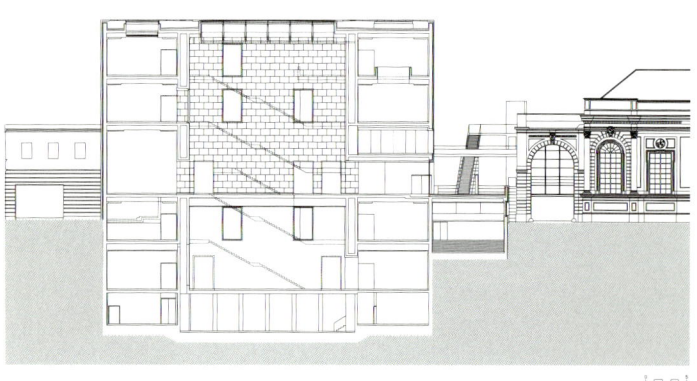

40m x 46m x 24m 크기의 콤팩트한 육면체로서 흰 대리석으로 완전히 덮인 이 박물관 건물은 그 주변과 완벽하게 일체를 이룬다. 석조 외피는 세심하게 처리된 섬세한 표면이며 파사드와 지붕은 모듈식 개구로 천공되어 있다.

10m 폭의 옥외 계단은 출입구로부터 시작되어 안뜰 위로 3.4m까지 올라간다. 내부로 들어서면 250m²의 유리 지붕 아트리움이 그 주위에 위치한 전시홀들을 위한 중심축이 된다. 아트리움의 높이는 19m이며 위의 3개 층을 연결한다. 천창을 통해 상부 아트리움과 연결된 하부 아트리움은 높이 9m로 아래 2개 층의 전시장을 연결한다.

긴 측면에 위치한 주계단실은 진입 레벨에서 샵과 카페가 들어선 중2층으로 이어지며 상부 2개 층과 하부 2개 층을 연결한다. 지하에는 창고와 서비스 시설로 이용되며, 2개 층 높이의 실 두 개가 아트리움에 인접해 있다.

보육 및 교육 시설은 기존 건물에 연결된 다리 내에 위치된다. 별도의 옥외 출입구를 가진 강당은 옥외 계단실 아래에 있다.

인접한 구 건물 윙에 위치한 행정실은 연결 다리를 통해 접근할 수 있다. 전시물 출하 구역과 작업실은 반원형의 윙 내에 위치하며 지하 터널을 통해 새 건물에 연결된다.

각 층에는 4개의 커다란 전시홀이 모듈식으로 배치되어 있으며 공동 접근 코어에 의해 서로 분리된다. 각 홀은 전시 개념에 따라 칸막이 벽으로 다시 분할된다. 일광은 동일한 크기의 창호를 통해 측면 또는 상부로부터 4개 층에 유입된다. 또한, 3개의 홀은 벽의 상부 가장자리를 따라 설치된 긴 창호를 통해 빛을 받아들인다. 일광이 전혀 들어오지 않는 박물관의 가장 낮은 층은 특별 전시실로 사용된다. 각 전시홀의 조명 시스템은 전시장 벽면을 균일한 조도로 만들며, 특수 조명들이 천장에 설치되어 있다.

재료: 파사드, 지붕, 벽, 천장 및 모든 방문객 동선의 바닥과 코어에는 흰 석회석^{부르고뉴사}이 사용되었으며, 전시홀의 바닥은 오크재 조각들로 구성되어 있다. 가시적인 모든 금속 부분에는 황동을 사용했다.

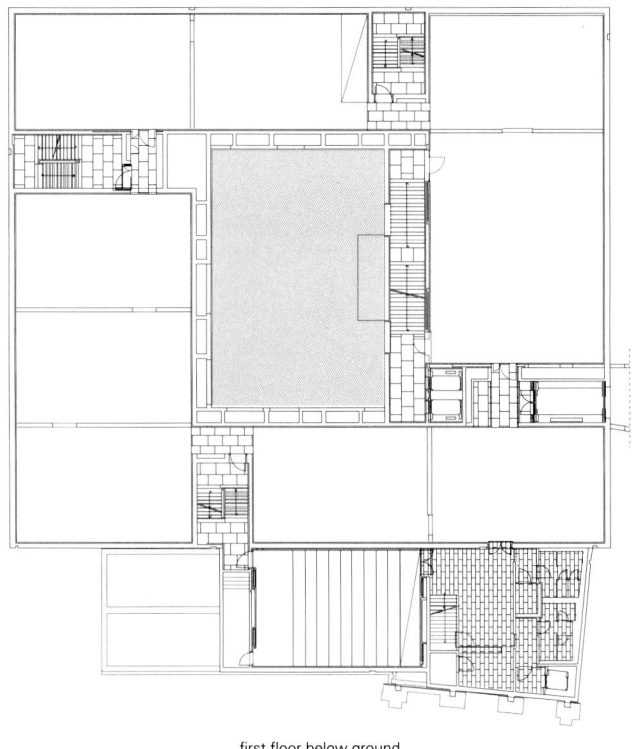
first floor below ground

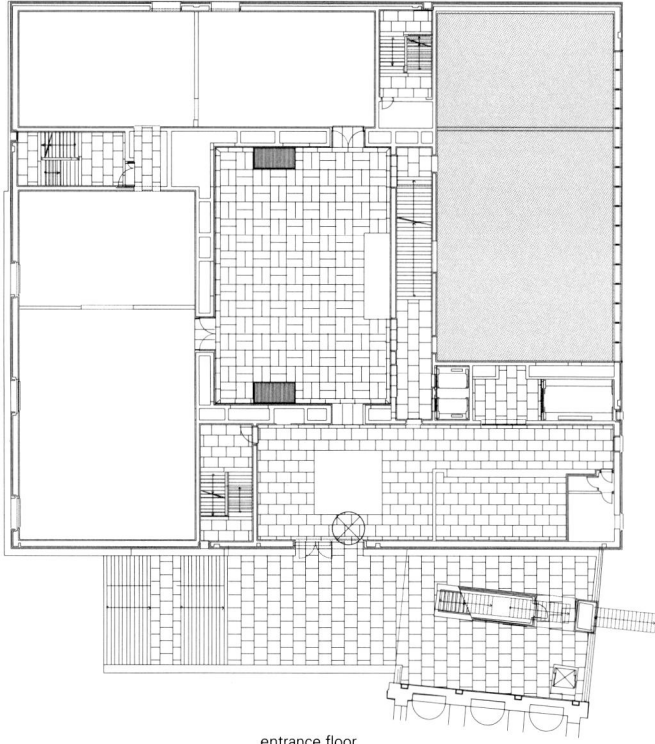
entrance floor

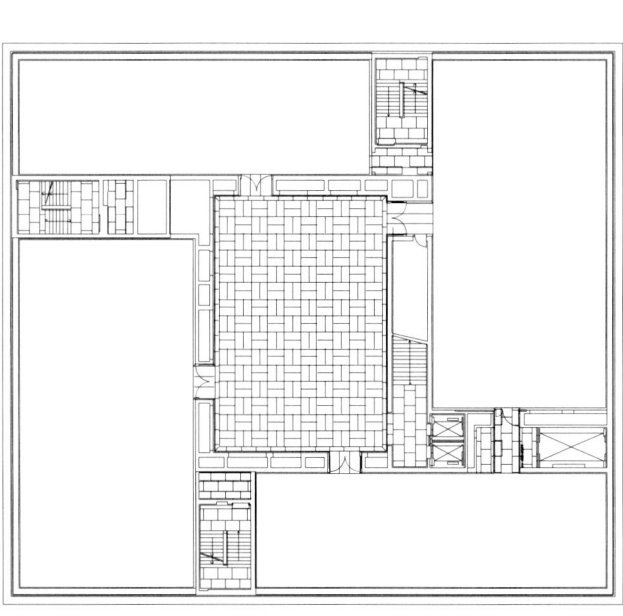
second floor below ground

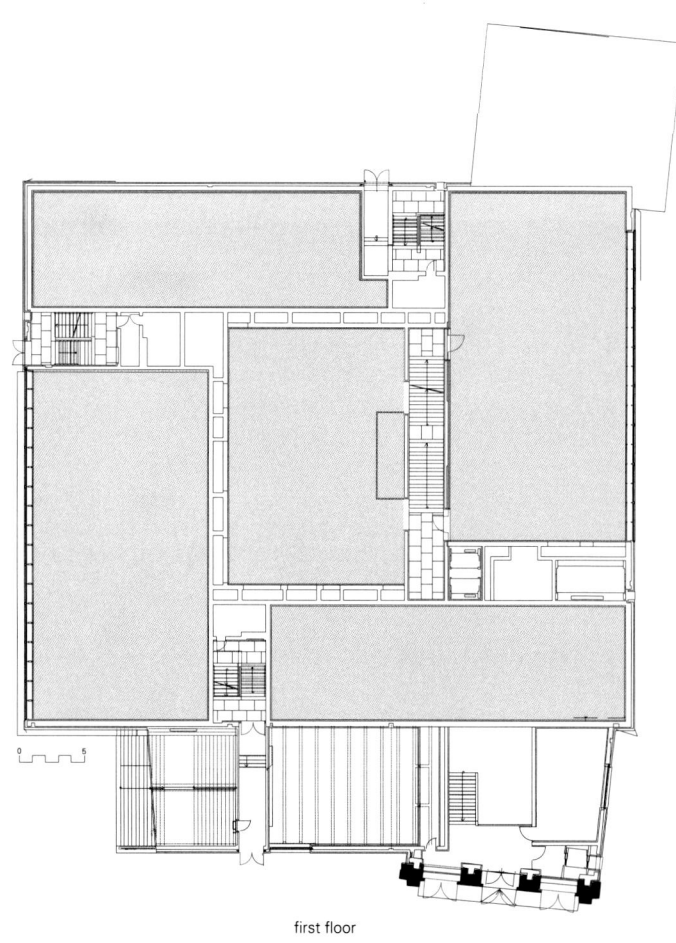
first floor

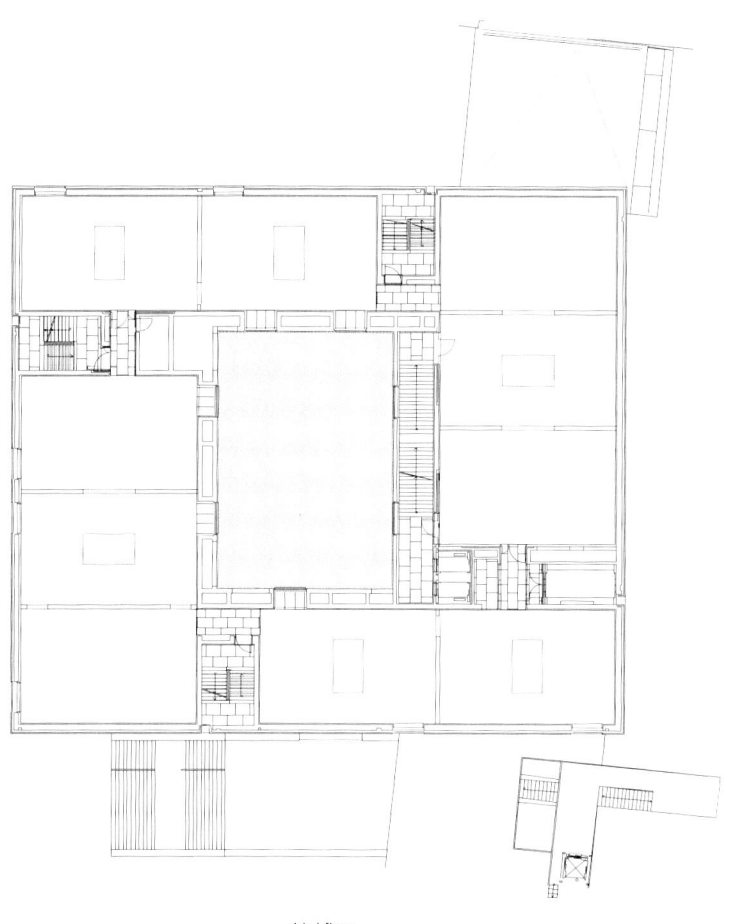

third floor

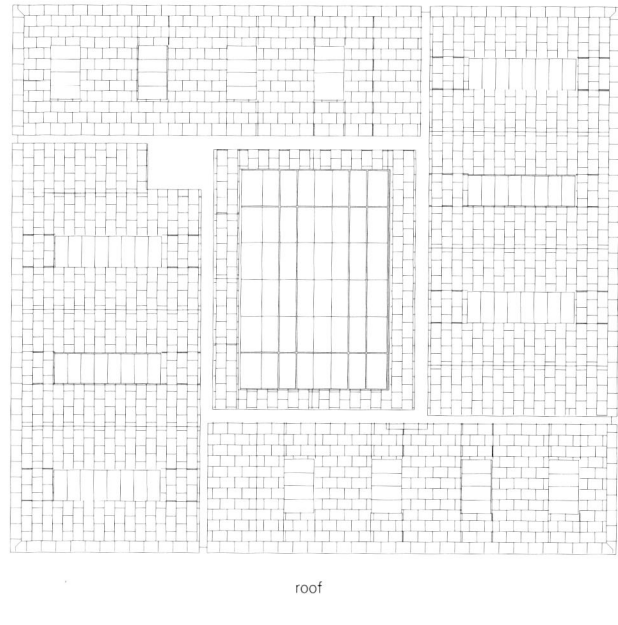

roof

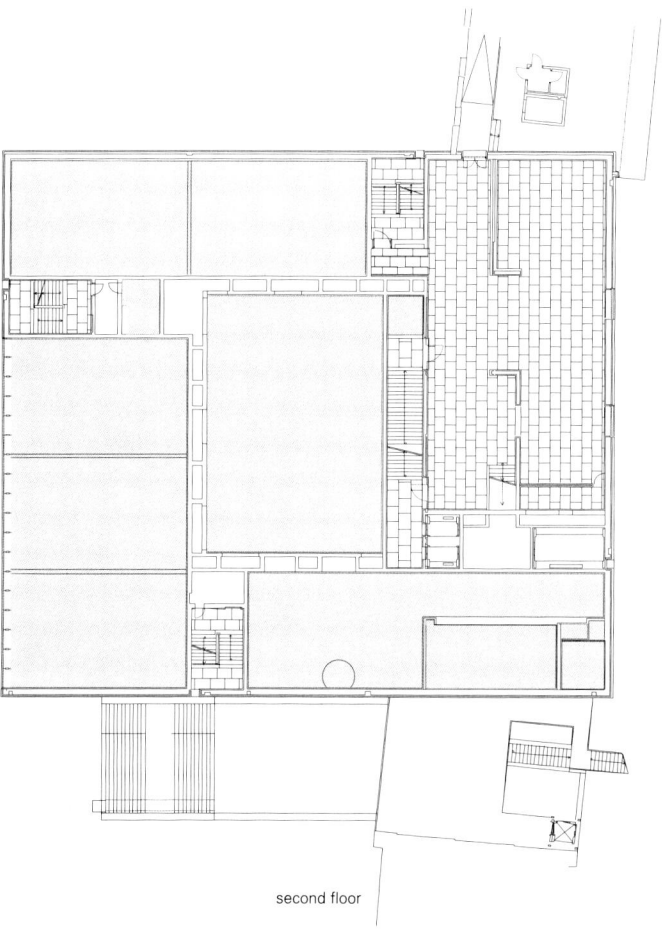

second floor

fourth floor

MUMOK
루드비히 근현대미술관

From the outside, the MUMOK ^{Museum Moderner Kunst Stiftung Ludwig Wien} appears like a dark, closed block. The roof curves down low on the edges. It is monolithically clad in anthracite gray basalt lava on the facades and roof surfaces. The building is clearly set apart from the adjacent level and seems to emerge from the deep.

A ten-meter-high outdoor stairway leads to the entrance level four meters above the courtyard level. Inside, a hall lit from above divides all of the levels into two differently proportioned groups of rooms. The entrance level is in the center of the building's height. Two main exhibition levels are above it and two below it. Another lower level is used for storage and utilities. On one side of access hall, five five-meter-high, pillar-free exhibition areas measuring about 700m² each are stacked above each other. These areas can be flexibly subdivided. On the other side, there are more intimate rooms measuring 250m² each. Here the ceilings are 3.5m high.

In between, openly positioned in the 35-meter-high hall, are the passenger elevator bank and the freight elevator. The various levels are connected with footbridges.

The first major exhibition level, the shop, and the cafe on the mezzanine connecting to the old building are directly adjacent to the foyer. An independently usable event area is below the outdoor stairway. The administrative offices are in the neighboring old building wing, and can be reached over the bridge.

Delivery areas and workshops are in the Oval Wing and are connected to the new building with an underground tunnel.

In terms of architectural design, the museum facilities are limited to generous reduction. They are equipped with a sophisticated and flexible artificial lighting system. The upper exhibition hall receives natural light through a large opening in the curved ceiling. The other slit-like openings and the panorama window on the uppermost floor give visitors a view to the outside and help provide a sense of purpose of orientation.

Materials Used : Basalt lava for the facade, roof, walls, and floors in the hall ; cast iron for the footbridges, steps, and wall paneling in the access core of the hall ; glass for railings and barriers ; terrazzo for the floors in the exhibition halls and all other public areas.

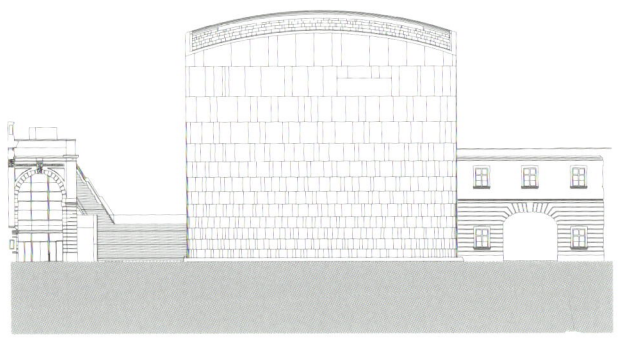
east elevation

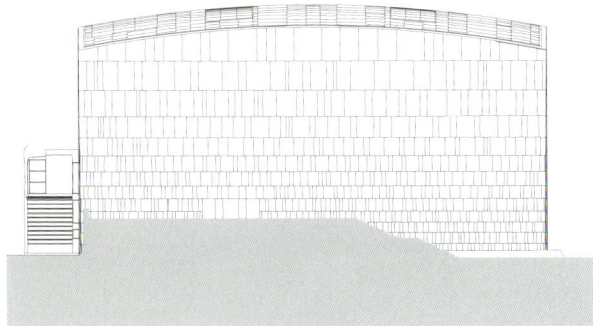
south elevation

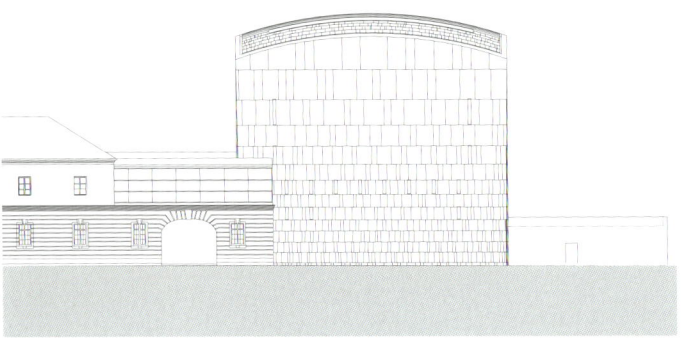
west elevation

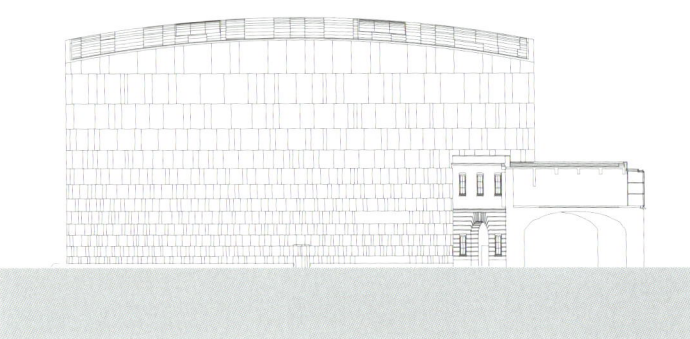
north elevation

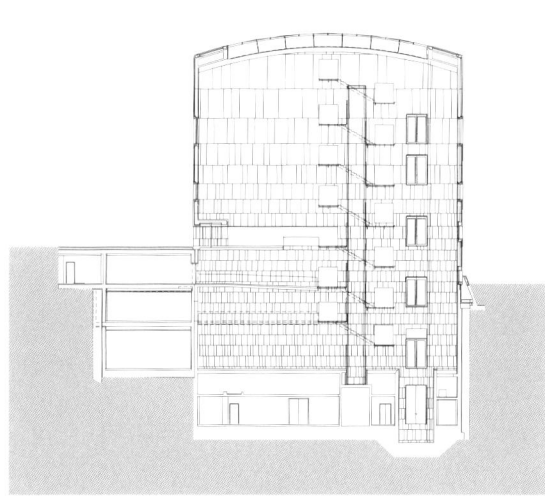
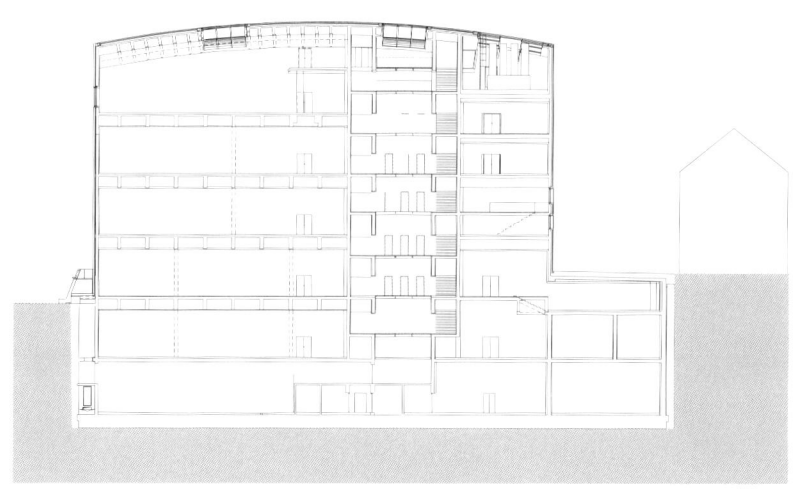

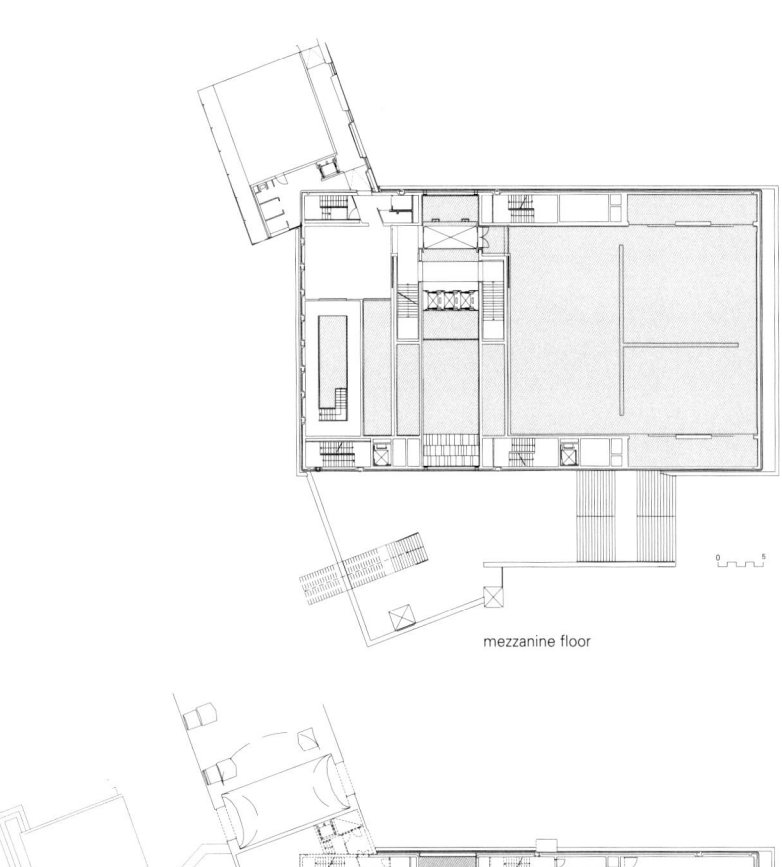
mezzanine floor

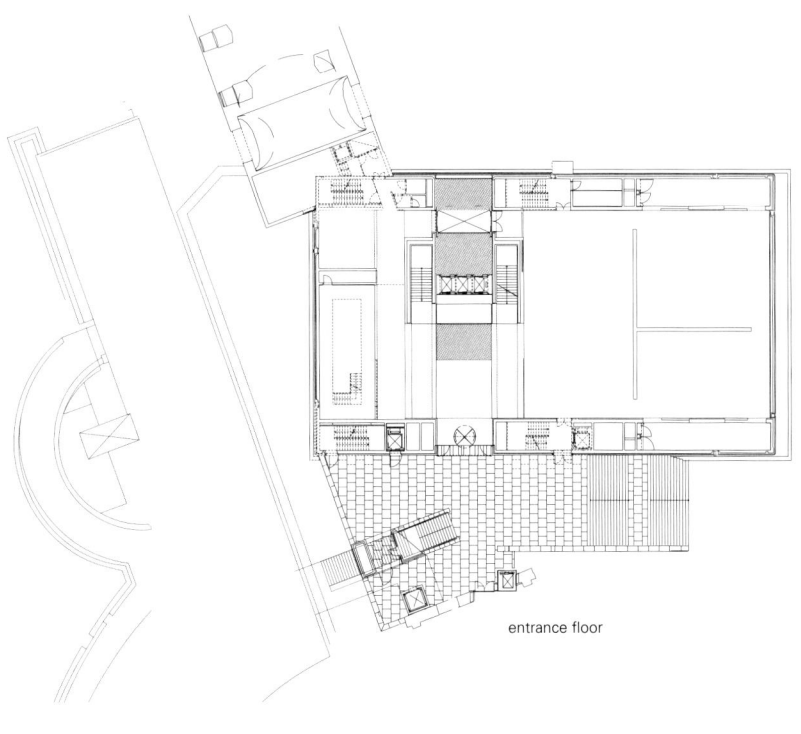
entrance floor

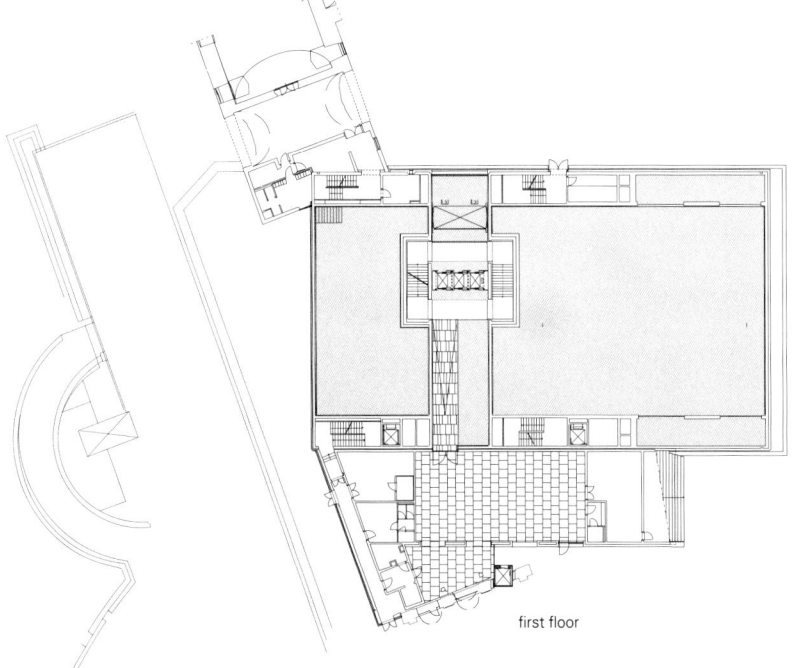
first floor

102 MUSEUM & EXHIBITION SPACE _ MUMOK

밖에서 보면 건물은 어둡고 폐쇄된 블록처럼 보인다. 지붕은 곡선을 이루면서 가장자리로 내려온다. 파사드와 지붕 표면은 모두 짙은 회색의 현무암으로 덮었다. 건물은 주변건물의 높이로부터 격리되어 깊은 땅속에서 솟아오른 것처럼 보인다.

10m 높이의 옥외 계단은 안뜰 위에 있는 4m 높이의 출입구로 이어진다. 안에서는 천창을 가진 중앙홀이 모든 층의 실을 다른 두 유형의 방으로 분할한다. 출입구는 입면상으로 건물의 중심에 있다. 두 개의 주 전시층이 각각 출입구의 위와 아래에 있고 지하 2층은 창고 및 서비스 시설을 수용한다. 출입홀의 한쪽 측면에는 기둥이 없는 5m 높이의 전시실들이 있는데 각 면적이 약 700m²인 이들 5개 전시실들은 상하로 위치하고 있다. 이들 구역은 유연하게 재분할될 수 있다. 다른쪽 측면에는 250m² 규모에 천장고 3.5m의 보다 친숙한 분위기의 방들이 들어서 있다.

35m 높이까지 오픈되어 있는 홀 사이에는 승객용 엘리베이터와 화물용 엘리베이터가 있고 다양한 층들은 홀을 중심으로 브릿지로 연결되어 있다.

기존 건물에 연결된 중 2층 내의 주요 전시 레벨과 샵, 그리고 카페는 로비에 인접하여 위치한다. 독립적으로 사용 가능한 행사장은 옥외 계단 아래에 있다. 행정실은 이웃한 기존 건물의 윙 내에 있으며 다리를 통해 접근할 수 있다.

전시물 인수/인출 구역과 작업실은 타원형의 윙 내에 있으며 지하 터널을 통해 새 건물과 연결된다.

디자인 측면에서 박물관 시설들은 가능한 최소화했고, 정교하며 유연한 인공조명 시스템을 갖추고 있다.

상부 전시실은 곡면 지붕에 뚫린 커다란 개구를 통해 일광을 받아들인다. 나머지 가늘고 긴 개구들과 최상층의 전망창을 통해 밖의 경치를 감상할 수 있고 동시에 방향 감각을 얻을 수 있다.

재료 : 홀의 파사드, 지붕, 벽과 바닥은 현무암을, 연결 브릿지, 계단, 그리고 홀의 출입 코어 내의 벽 판넬에는 주철을 사용했다. 난간과 칸막이는 유리로 처리했고 전시홀 및 다른 공공구역의 바닥에는 테라조가 사용되었다.

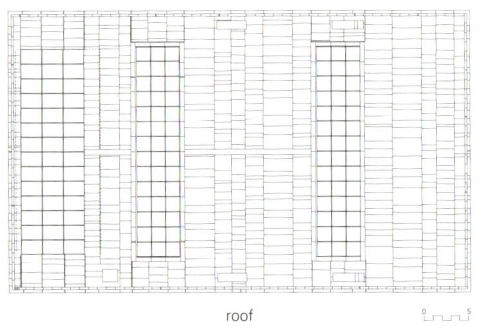
roof

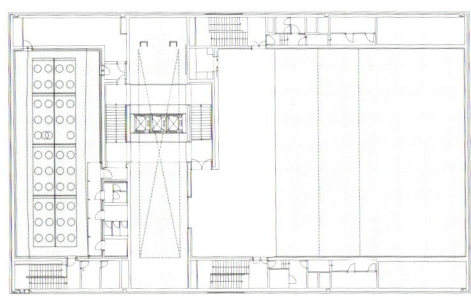
fourth floor

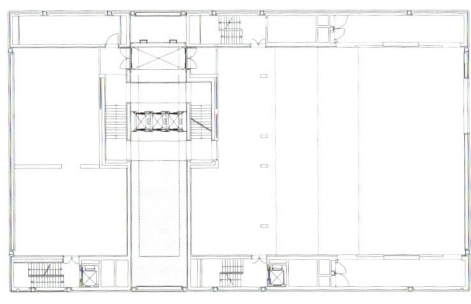
third floor

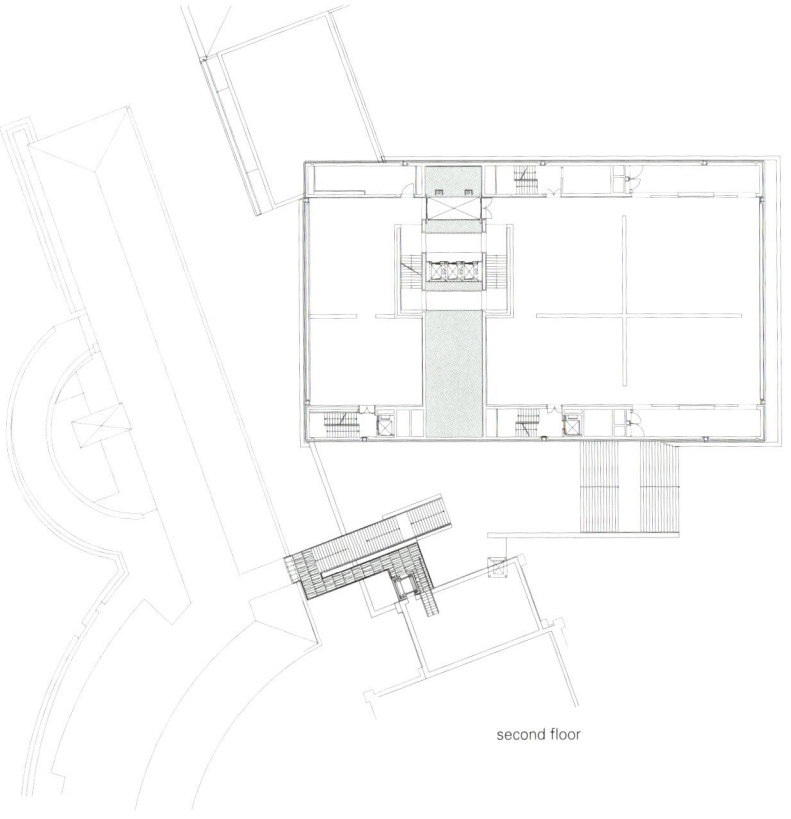
second floor

104 MUSEUM & EXHIBITION SPACE _ MUMOK

Miho Museum
미호미술관

I. M. Pei
I. M. 페이

The religious sect of Shinji Shumeikai is devoted to the ceaseless pursuit of beauty, both as expressed in nature and also in the works of art which the sect, under its spiritual leader, avidly collects.

The Miho Museum combines both aims - nature and art - in a 17,400m² building (82,865 square feet) housing one of the world's finest ancient Japanese and early 'Western' collections. The building itself is one with nature; its asymmetrical footprint is determined by the contours of the irregular site, its materials derive from natural sources, and the bulk of the building? Nearly 80% - is recessed into the ground so that its wooded mountain setting remains substantially intact. Approaches to the museum are configured through peaks and over valleys in order to preserve the terrain unscarred by open cut roads.

The clairvoyant Kaishusama, spiritual leader of Shinji Shumeikai, selected the site between twin ridges of a forest preserve about one kilometer east of the sect's sacred precinct. But unlike the remote private sanctuary, Miho has been undertaken as a public foundation with a projected public attendance of 70,000 annually. Visitors arrive from Kyoto by car or bus at a triangular Reception Pavilion articulated on the exterior with Japanese stucco and tile and housing a cafeteria and ticketing services within.

From here guests are transported on small electric cars through one of two new tunnels cut through the mountains to reach the museum(The other, a 900m/.6 mile service tunnel dedicated to art transport and emergency evacuation, is twice as long, curving around topographical depressions in order to remain underground). The mouth of the visitor tunnel opens onto a 120m (400 foot) suspension bridge which spans a precipitous drop to land visitors directly onto the museum plaza. The approach is a measured procession into Shangri- la: as visitors emerge from the tunnel onto the spiderweb bridge, they glimpse their first, partial views of the museum in its undisturbed mountain paradise. There is little sense of the building's scope as it appears like a series of hipped skylights resting on the earth. Vertical elevations are for the most part expressed on the opposite/west facade where, following the site's folded contours, they never exceed a height of 13m (43 feet) above grade.

From a circular drop-off visitors ascend a series of leveling terraces in the fashion of a Japanese temple and enter the museum's main public space: a space framed reception hall with innovative skylight and aluminum sunscreens. In an advance over less durable wood veneers, all surfaces have been digitized to evoke the grain, texture and warmth of local woods, thus merging technology and tradition, western and eastern cultures, interior and exterior space.

Through 7m(23-feet) high transparent window walls the hall embraces the valley sprawled to the west, overlooking the sacred precinct with its emblematic bell tower and sanctuary. From the reception hall visitors may proceed to the museum's north or south wings, both 1,000m² but varied in form and position to accommodate the land. The north wing is devoted to Japanese art and is organized around a square Japanese garden left open to the sky. Housed below are conservation laboratories, storage, and service facilities. The curved south wing, by contrast, is three stories high and devoted to 'western' art and museum operations. The top level is occupied by administration and also by Egyptian art galleries. Exhibition areas continue below with rich installations of Gandharan, Near Eastern, Greek / Roman, Ancient Chinese, Sassanian, and Islamic art. An auditorium and public tea room with and open courtyard are also located on this main level of the museum. The bottom floor of the south wing houses private curatorial space.

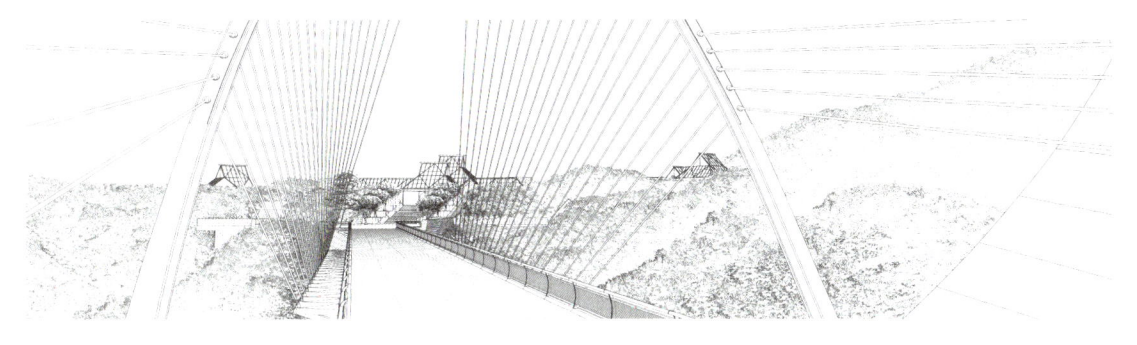

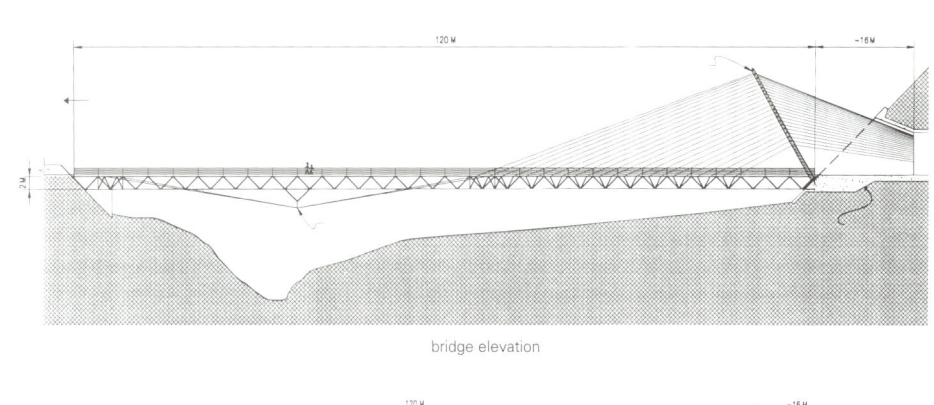

bridge elevation

roadway plan

space frame plan

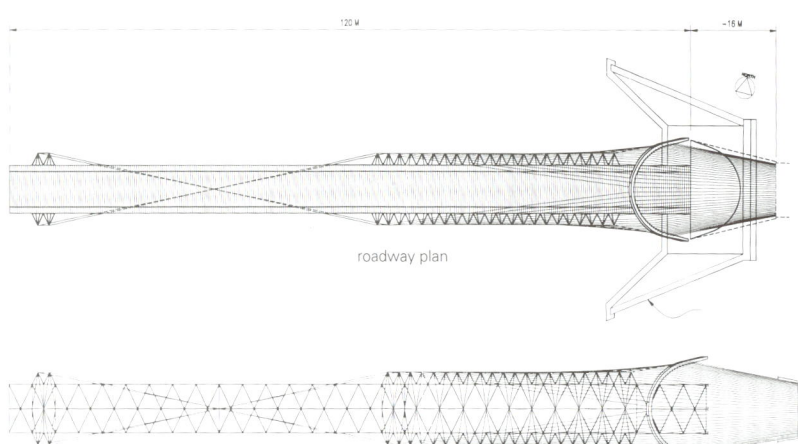

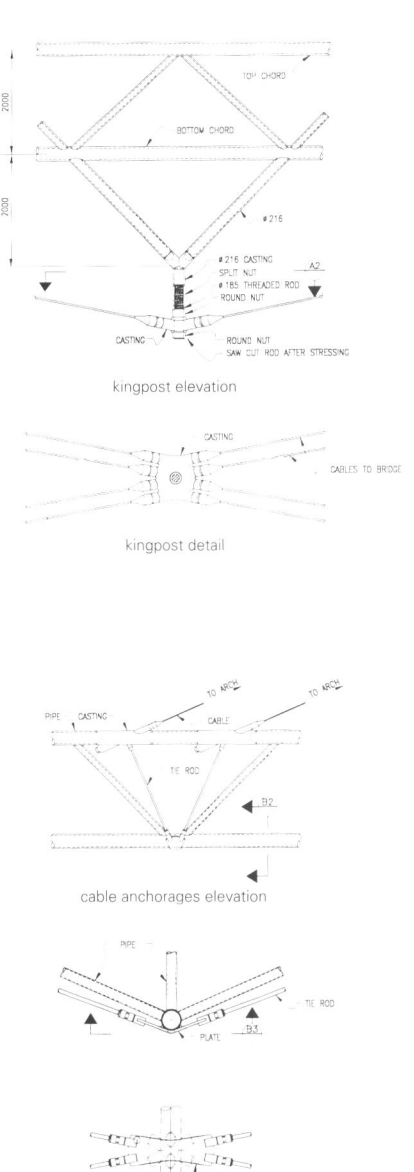

kingpost elevation

kingpost detail

cable anchorages elevation

bottom chord detail

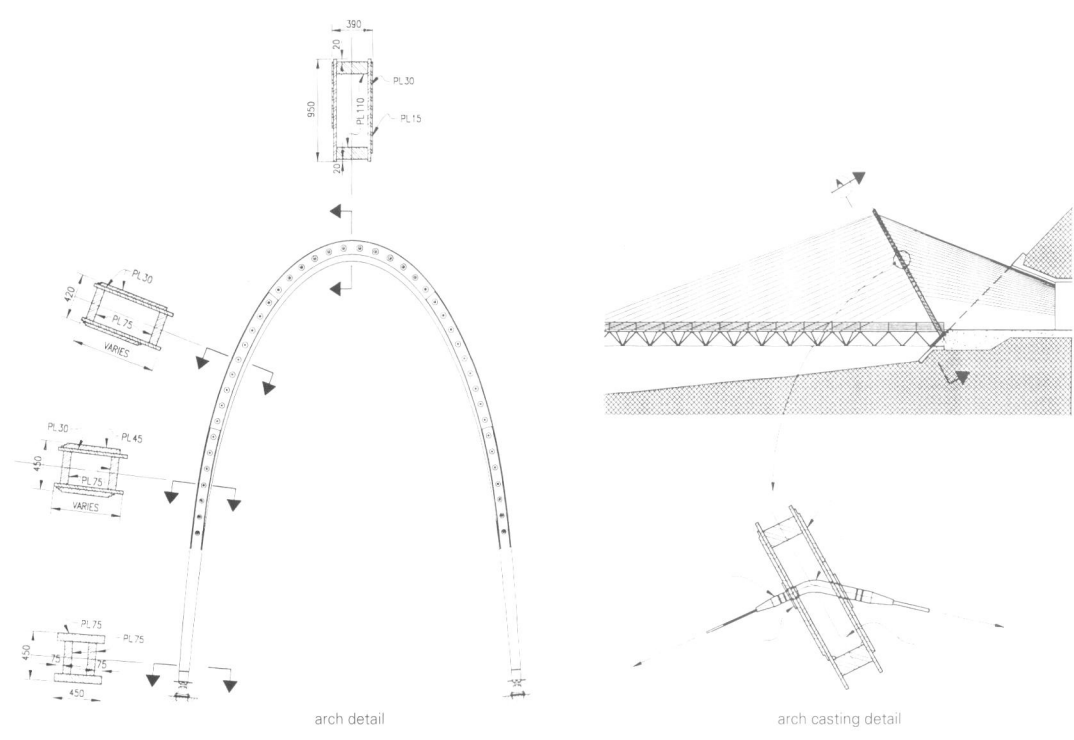

arch detail

arch casting detail

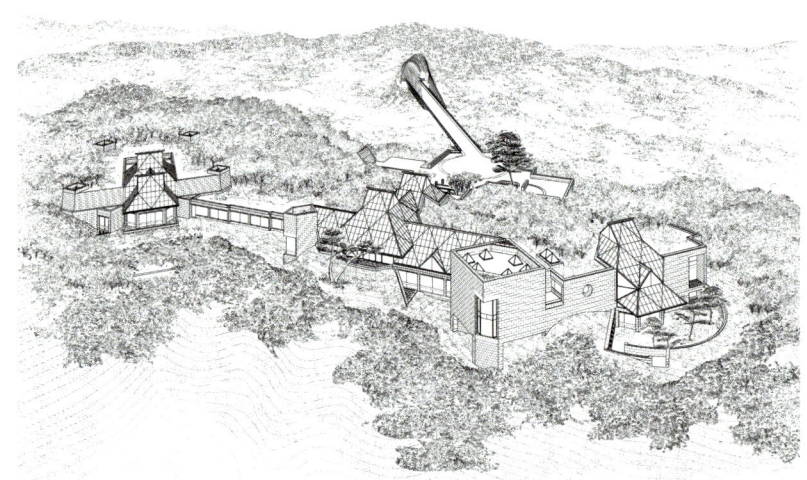

신지 슈메이카이 종파는 자연과 예술작품 속에 표현된 미의 추구를 그들 최고의 미덕으로 삼고 있다. 그들이 자신들의 정신적인 지도자의 지도 아래 예술작품 수집에 열을 올리는 것도 이런 맥락에서이다.

미호미술관은 자연과 예술이라는 신지 슈메이카이 종파의 두 가지 목적을 결합하여 17,400m^2의 건물 내에 전세계에서 가장 훌륭한 작품 중의 하나인 일본 고대 미술품과 초기 서구 미술품들을 소장하고 있다. 건물 자체는 마치 자연의 일부처럼 보인다. 천연자재를 활용한 비대칭의 건물형태는 불규칙한 대지윤곽에 의해 결정되어졌으며 건물의 80% 정도가 대지에 푹 파묻혀 있어 수목이 우거진 주변환경이 거의 그대로 보전되었다. 도로를 내게 되면 수변환경이 훼손될 것을 우려해 계곡과 산을 거쳐 미술관에 출입하도록 했다.

신지 슈메이카이의 정신적 지도자로시 통찰력이 뛰어난 카이슈사마는 신지 슈메이카이의 성지에서 약 1km 동쪽에 위치한 삼림보호구역의 쌍동이봉 사이를 대지로 선정했다. 외딴 곳에 있는 다른 성소들과 달리 미호미술관은 연 7만여 명의 관람객이 모여들 것으로 예상, 그에 맞는 공공건물의 개념으로 구상되었다.

방문객들은 교토에서 승용차나 버스를 이용해 간이식당 및 매표소가 있는 삼각형의 출입구 별관에 이르게 된다. 일본 특유의 치장벽토와 기와가 외장재로 사용된 이 출입구 별관에서부터 아담한 전기차를 타고 산중턱에 신축한 터널을 지나면 미술관에 도착하게 된다. 미술품 운송 및 긴급대피용인 또 다른 터널은 신축 터널의 두 배인 900m 길이로 지상으로 나오지 않기 위해 함몰된 지반 주위를 곡선형으로 선회하게 되어 있다. 방문객 터널의 출구는 120m 길이의 현수교로 이어지고 현수교를 통해 가파른 절벽을 지나면 미술관 광장이 바로 나온다. 이러한 진행방향은 철저하게 계산된 것으로 방문객들은 어두컴컴한 터널을 지나 흡사 거미줄 같은 현수교로 나와 쏟아지는 빛에 눈을 깜박이면서 훼손되지 않은 자연 한가운데 놓여 있는 미술관의 전경을 보게 된다. 말 그대로 지상낙원에 도달하는 것이다. 이런 맥락에서 건물의 규모는 별 의미가 없다. 추녀마루가 있는 지붕들이 땅 위에 연속적으로 늘어서 있을 뿐이다. 서측파사드에는 수직입면이 뚜렷하게 표현되어 있는데 이쪽에서는 중첩되는 지형에 따라 입면이 지면에서 13m를 초과하지 않는다.

방문객들은 나선형 절벽으로부터 일본 사원에서 흔히 찾아볼 수 있는 여러 층에 걸친 테라스들을 지나 미술관의 주요 공공공간인 이색적인 천창과 알루미늄 차양이 있는 리셉션 홀에 들어간다. 내구성이 떨어지는 합판 대신 표면 전체를 디지털화하고, 이 지역에서 나는 원목 특유의 질감과 온화함을 이용해 기술과 전통, 서구문화와 동양문화, 내부공간과 외부공간을 접목시켰다.

리셉션 홀은 7m 높이의 투명한 창으로 구성된 벽들을 통해 서쪽에 펼쳐진 계곡을 포용하는 동시에 종탑과 성소가 딸린 성역을 굽어보고 있다. 방문객들은 리셉션 홀에서 북측이나 남측 윙부분으로 갈 수 있는데, 이 두 부분의 면적은 동일하지만 형태 및 위치는 상이하다.

일본 미술품만을 전시하는 북측 윙은 하늘을 향해 열린 구조를 취하고 있는 정사각형의 일본식 정원 주위에 자리잡고 있다. 그 아래쪽에는 문화재보호연구소, 창고, 기계실 등이 있다. 이와 대조적으로 3층 높이의 곡선형 남측 윙에는 서구미술품이 전시되어 있으며, 미술관 운영사무실이 있다. 최상층에는 행정사무실 및 이집트 미술 전시관이 있다. 아래층에는 간다라, 근동, 그리스·로마, 고대 중국, 사산 왕조, 회교 미술품들이 다채롭게 전시되어 있다. 미술관의 주 레벨인 이곳에는 개방된 중정이 딸린 다실과 강당도 마련되어 있다. 남측 윙의 맨 아래층에는 큐레이터 사무실이 있다.

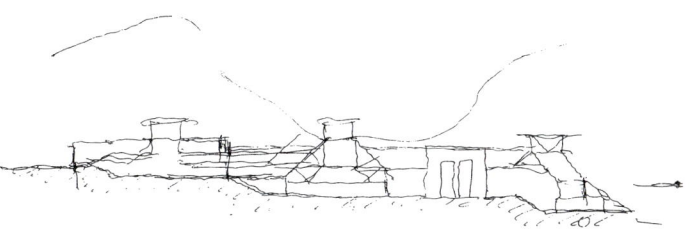

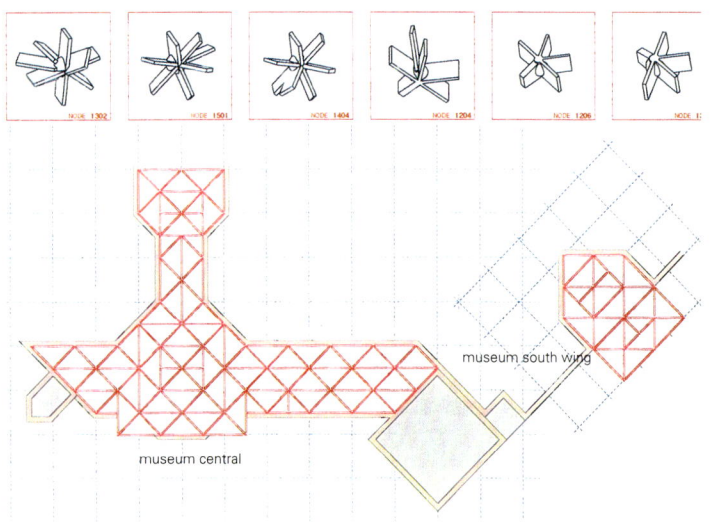

museum south wing

museum central

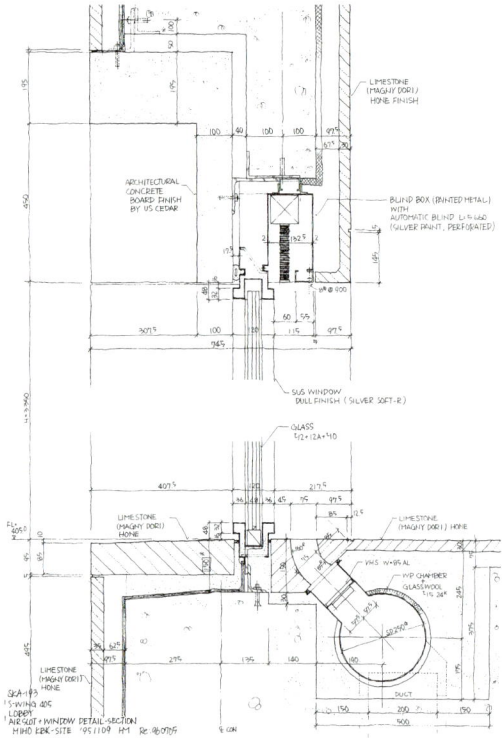

south wing lobby air slot and window section detail

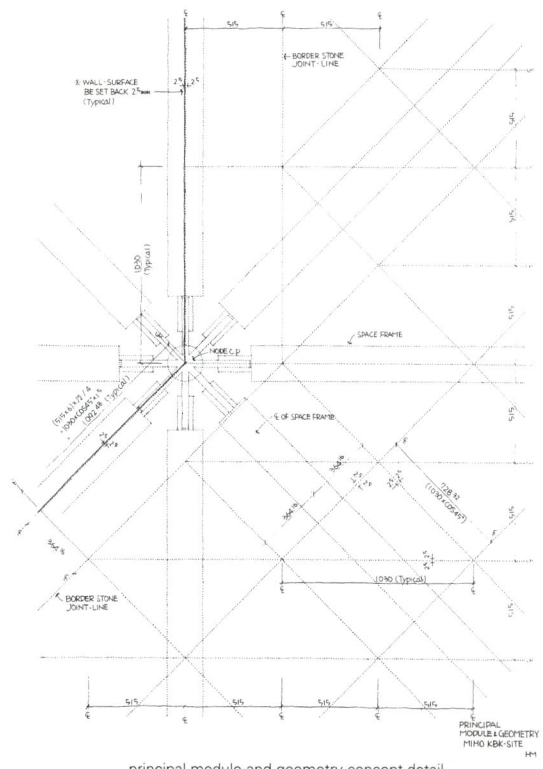

principal module and geometry concept detail

Architect: I.M. Pei Architect
Project architect/Tim Culbert
Building envelope/Perry Chin
Initial phase design/Chris Rand
Project team/Carol Averill, Price Harrison,
Celia Imrey, Hubert Poole
Associate architect: Kibowkan International Inc.
Principal manager in charge/Osamu Sato
Principal administration/Hiroyasu Toyokawa
Project team/Fumio Ozaki, Hitoshi Maehara,
Masa Sato, Yasuhiro Sonoki, Miho Toyoda
Landscape architect: Special & gardens project team/
Kohseki and Akenuki Zone-Y. Nakamura, A. Akenuki,
T. Culbert
General site landscaping/Noda Kensetsu, Takuji Noda
Consulting engineers: Structural/Leslie E. Robertson
Associates-Leslie E. Robertson, Saw Teen See,
Katherine Hill
Lighting/Fisher Marantz Renfroe Stone-Paul
Marantz, Alicia Kapheim, Hank Forrest
General contractor: Shimizu Corporation-Mutsuo
Takagi, Mikio Noguchi
Client: Shinji Shumeikai
Location: Shigaraki, Shiga Prefecture, Japan
Site area: 1,002,000m^2
Gross building area: Museum/17,429m^2
Reception pavilion/3,351m^2
Use: Museum for ancient Japanese and Western Art
Photograph: Timothy Hursley

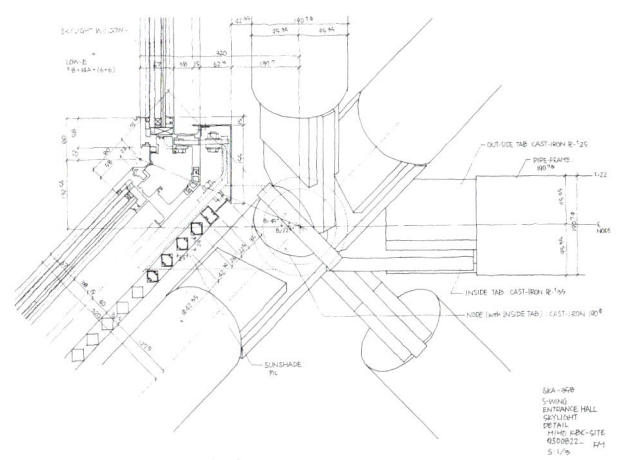

south wing entrance hall skylight detail

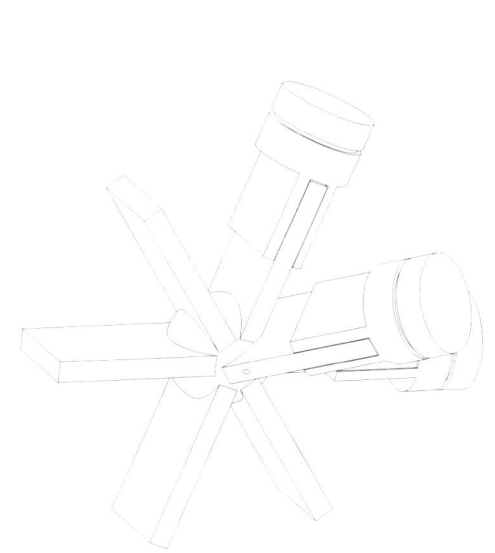

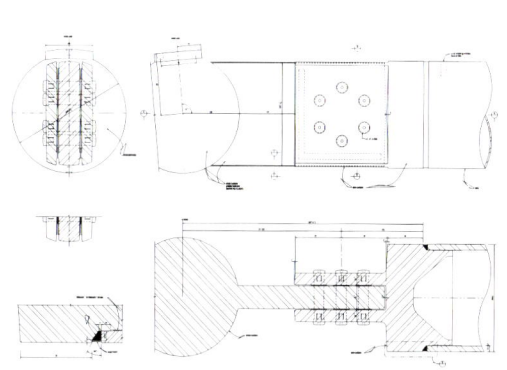

typical pipe node detail

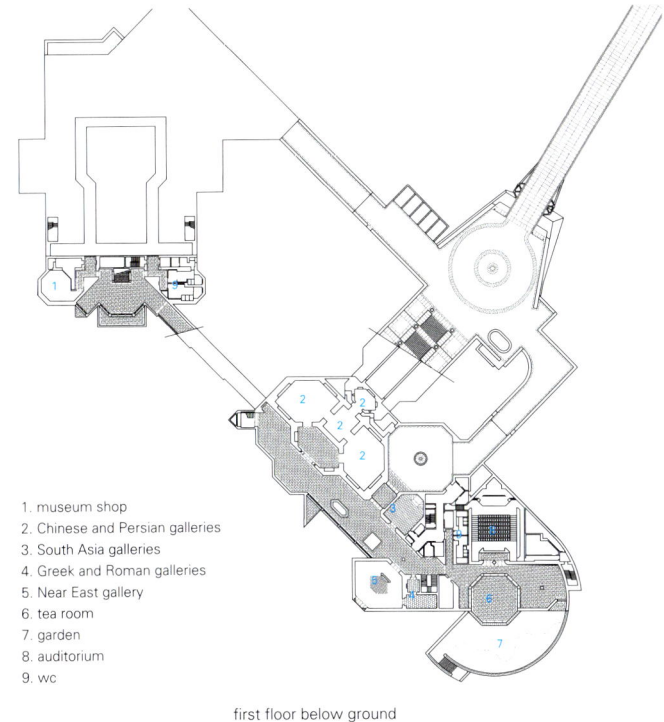

1. museum shop
2. Chinese and Persian galleries
3. South Asia galleries
4. Greek and Roman galleries
5. Near East gallery
6. tea room
7. garden
8. auditorium
9. wc

first floor below ground

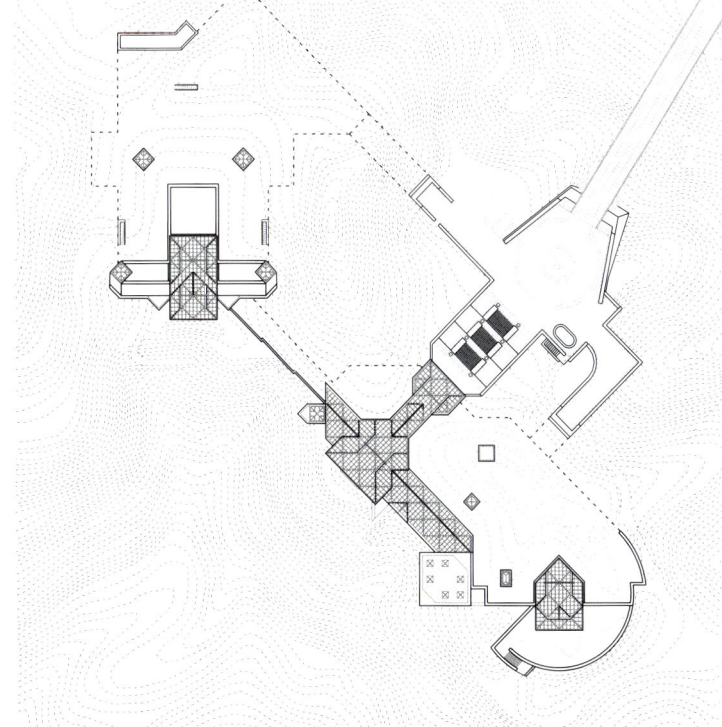

roof level

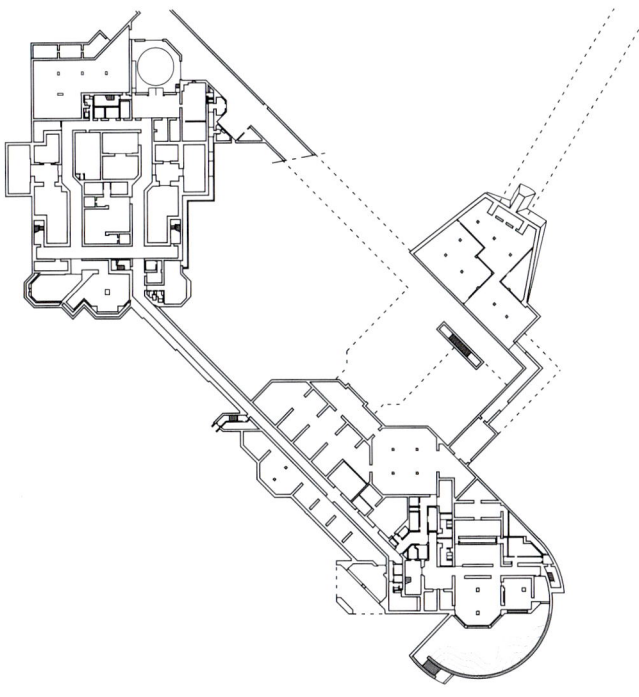

second floor below ground

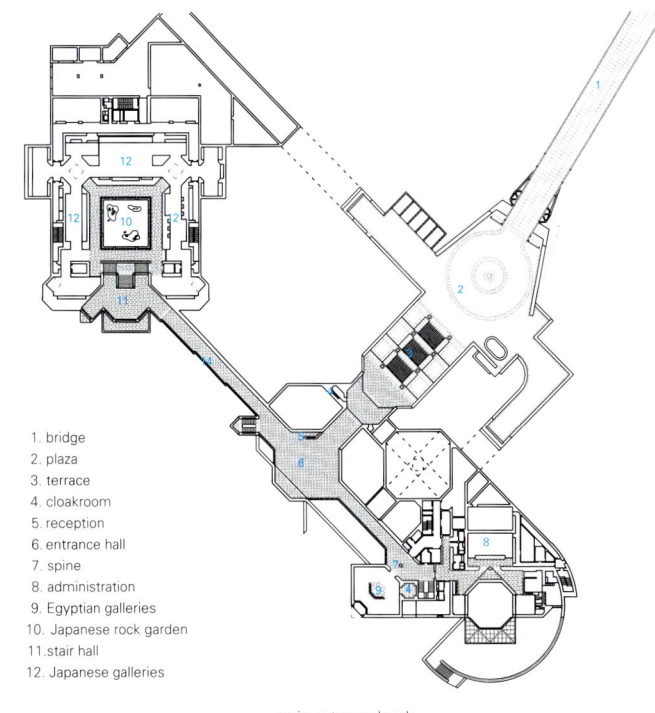

1. bridge
2. plaza
3. terrace
4. cloakroom
5. reception
6. entrance hall
7. spine
8. administration
9. Egyptian galleries
10. Japanese rock garden
11. stair hall
12. Japanese galleries

main entrance level

Lille Fine Arts Museum
릴 미술관

Jean - Marc Ibos & Myrto Vitart
이보스 앤 비따르

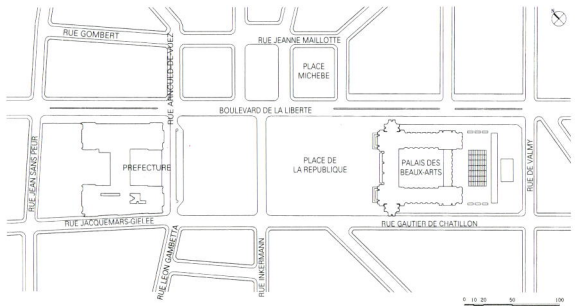

The most curious phenomenon today, and for us the most interesting, consist in the development of cities. Or to be more precise, how the contemporary city avoids all attempts for global mastering. Each time there is a temptation of mastering the city as planified, organized, canalised entity, the result is extremely poor. It is an abstraction of the city, a sliced city that is proposed to us.
Every one is obviously surprised in front of phenomenon which only tangible nature might be indefinable.
The classical city is a graspable entity, hierarchically controlled and organized, image of a society that itself was hierarchically controlled and organized.
In these cities, the whole has a meaning.
The ancient city is thought according to an absolute, transcendental and irreducible conception of beauty. We are indeed sensitive to it.
We are also conscious of the fact that, today, these cities are no longer the sole representation of our diversity and our complexity.
It is illusory to make today's world fit in such a vision.
If, constrained, things reappear, unexpectedly.
Therefore, we prefer another approach ; things would have an inner value and it is in that value that would lie their beauty, directly connected to reality. We like to think that beauty could be all relative, linked to the essence of things, and to the way look at them. This point of view allows infinite possibilities. We can then consider the renewal in a positive way.
Enhancing existing qualities constitute our strategy. Utilizing potentials permits an economy of means. We don't pain trying to bring life to abstractions, we use life. We target the effort to put in, according to the places we work with.
We can then transform them, integrate them, play with them easily.
The will of lightness, of flexibility in the way we approach things leads to conceptions open in terms of use, open in terms of interpretation. We want our designs to hold the necessary unspecified portion that allows them to integrate the real world.

The Fine Arts Museum built in 1892 is a reduced version of a larger initial proposal. The original Bérard and Delmas scheme was twice the size the actual construction, covering the entire lot up to the rue de Valmy.

In the thirties, the central courtyard is converted into an atrium, topped with a glass ceiling.This atrium becomes the center of the building, around wich the collections are organised. A monumental stairway leading to the painting galleries on the 2nd floor is then erected, thus blocking the south arcade from the inside.
In the seventies, it is the exterior south arcade wich is blocked by the addition of an administrative wing. Simultane-ously, the grand Wicar gallery, located in the north wing facing the place de la République is divided into 3 levels, two of wich are used for art storage.
The lack of space and the vetusty of the installations, contribute as well to the arbandonment of the building. In this context, the proposal to house part of the raised relief model collections from the Musée des Invalides, gives at the end of the eighties the opportunity to 're-think' the museum in its globality.
The 1990 competition programm requires the creation of new spaces devoted to the raised relief models, the creation of a tempory exhibit gallery, of an auditorium, of a library and of children work rooms as well as the reorganisation of the administrative services. It is also requested that the existing building is brought up to code and that the architects will propose a new concept for the presentation of the painting, the medieval, the ceramic and the 19th century sculpture collections.
The Lille Fine Arts Museum is set on a highly privileged position in the centre of the town. Facing the Prefecture, the two buildings of the same period, interrack across the Place de la République.
Built only half the size planned, it was later on subjected to several affronts which were demonstrated by an overall congestion of space. Galleries were sectionned lengthways with mezzanines, the facade on the garden was blocked up by an extension in the seventies. The very meaning of the project was corrupted.
This building is characteristic of a solemn style of architecture made up of bulk, shadow and light, of a luxury of space inconceivable today, of majectic volumes and superb successions of interior perspectives.
Failing to restore it to its normality, we were continually running the risk of being restricted by the place. We chose to go along with what existed and to make the most of its qualities. In some ways, the whole extension project stemmed from this guiding principle.

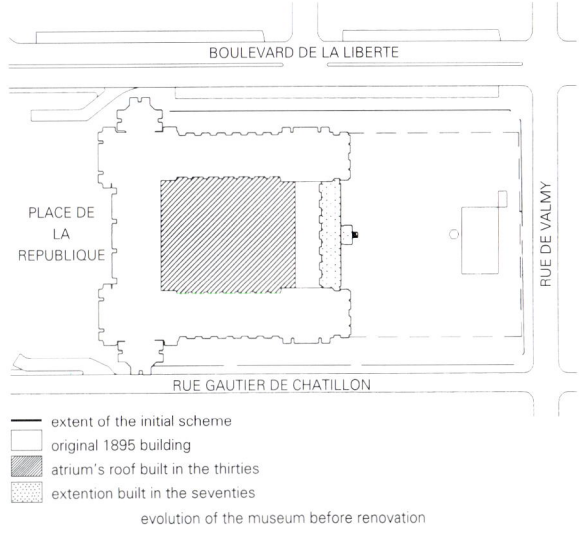

- extent of the initial scheme
- original 1895 building
- atrium's roof built in the thirties
- extention built in the seventies

evolution of the museum before renovation

Making the most of the privileged situation of the museum, we have opened it up to the town, convinced that a living museum is above all a museum open to its time. We have cleaned and stripped the museum of its dross. We have restored and increased the visual balance, the depths, the perspectives.

The extension of the seventies was demolished, as well as the central staircase in the atrium which cluttered up this space. The arcades were reopened.

The museum is literally projected to the outside with, as a background perspective, a fine blade which restitute the reading of the initial volume, never realised, of the Bérard and Delmas project in 1895.

Through this restitution, we integrate the outside space within the museum. The sculpture garden becomes a room open to the sky, next to the boulevard de la République. This 'blade' is made up of a series of vertical planes which are superimposed.

The first plane is made of clear glass which reflects the impressionist image of the Fine Arts Museum.

Beyond that plane are gold monochromes on a red background. The same gold and red that apear along the visit. All this make up the emblem of the museum.

The reflection is the interface between the old and the new. The relief of the 19th century building is reflected by the image of this relief. The new building is the supporting structure of this image. By emphasising the depths, the mirror registers the extension into the logic of the existing.

The entrance to the museum is through the two original doors.

Consistent with the logic of opening up which we have expressed, we propose a ground floor with free access from the Place de la République going through to the rue de Valmy, passing through the atrium, the sculpture garden and the café-restaurant.

The viewing of the collections is organised on 3 lavels around the atrium.

We propose a varied tour, alternating the spaces where we made slight alterations, perceptible but not directive, and areas of density which we have stressed most forcefully.

To accommodate these collections, we disposed of very different types of space which we have tried at all costs not to homogenise.

Here again, our attitude was to reveal each place according to its logic and to allocate the appropriate collections effectively.

The second floor which disposed of vast galleries with lighting from above remains allocated to painting. The red of the walls reconsiliate the artwork with the monumentaly of the space.

On the ground floor, the east and west galleries made of limestore, house the sculptures and the ceramics.

The Mediaeval and Renaissance section, in all its modes, is situated in the basement in the recuperated cellars which we have kept as they were. The high brick vaults have been cleaned : only minimal changes were made according to the logic of additions.

We are preserving the emptyness of the galleries, and the rythmic perseption of the brick arcs. The pieces, mostly religions are presented on massive oak stand, of wich we colored the top surface as a transition surface.

The space reserved for the raised relief models section was obtained by excavating under the atrium. We have thus recuperated a huge free uninterrupted space in keeping with the scale of the works. Here the walls, the ceiling and the floor are black. The lighting is precisely cut around the models and allow for an overall vision of the collection.

In two strategic places, we welcome contemporary artists whose work in situ materialises fully the idea of a living museum. Gaetano Pesce proposes in the entrance pavilions large centre lights which invade the space of the rotundas 'like two bags containing objects in fragments, witnesses of everyday reality'. Guilio Paolini with a series of 48 cubes in serigraphed glass has taken over the atrium to elaborate his idea that "it is our point of view and not the object (always equal or destined to become so), it is the trajectory of our expression (always different but nevertheless unique) which draws here or elsewhere the exhibition space, the place of the work, this theatre of light and silence which is a museum."

Taking into account the very essence of things and places which are given to us allow us to interpret them with the greatest latitude. Past and present can then coexist freely and mutually enrich each other.

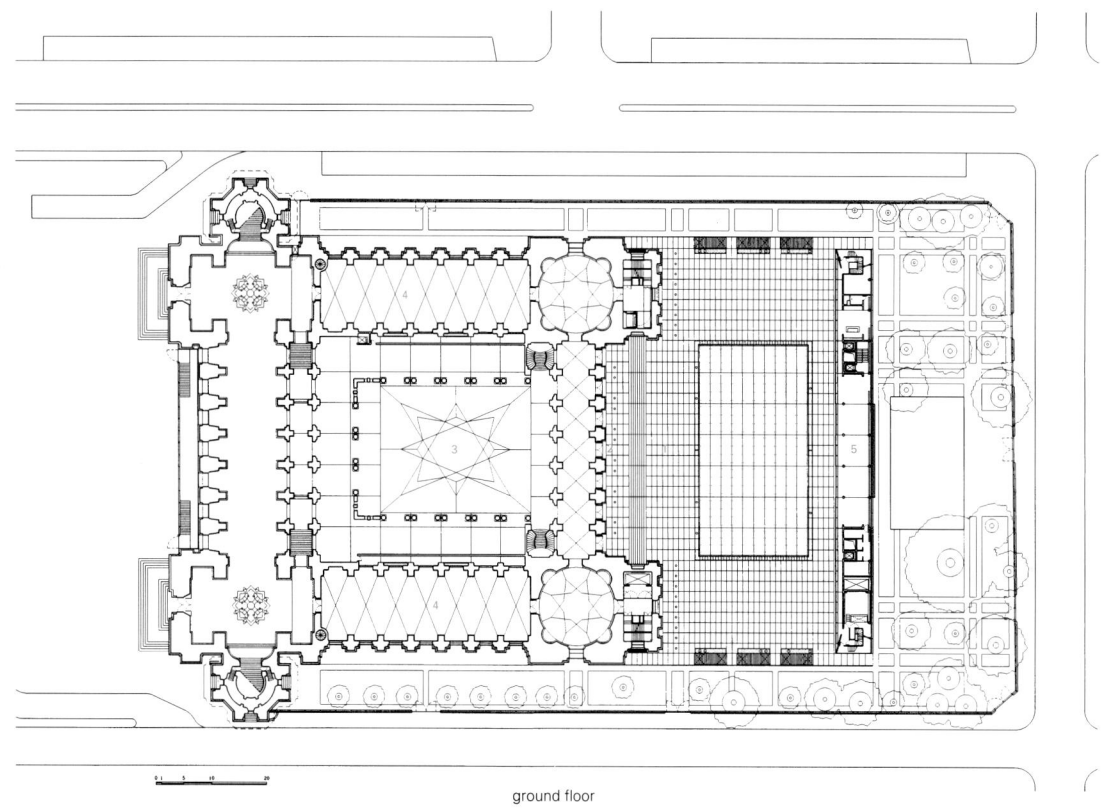

ground floor

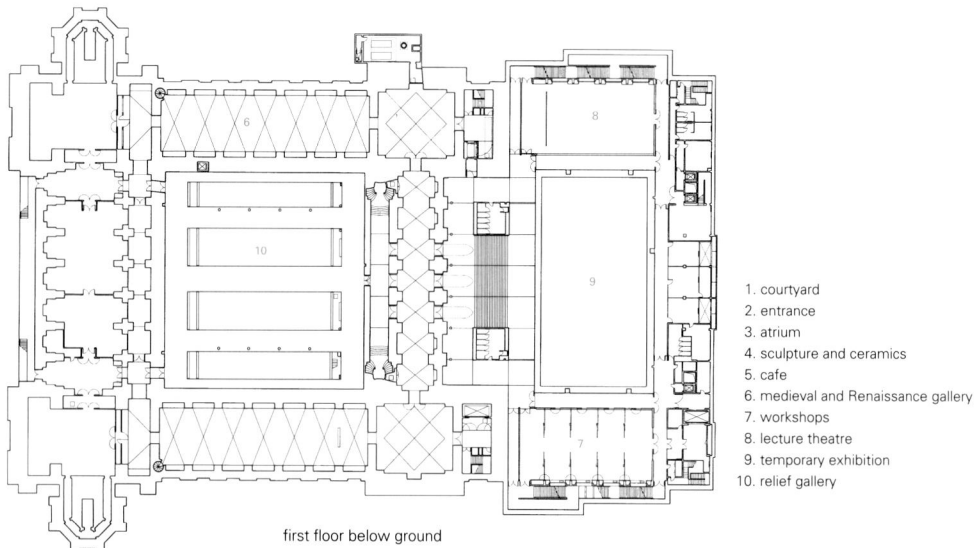

1. courtyard
2. entrance
3. atrium
4. sculpture and ceramics
5. cafe
6. medieval and Renaissance gallery
7. workshops
8. lecture theatre
9. temporary exhibition
10. relief gallery

first floor below ground

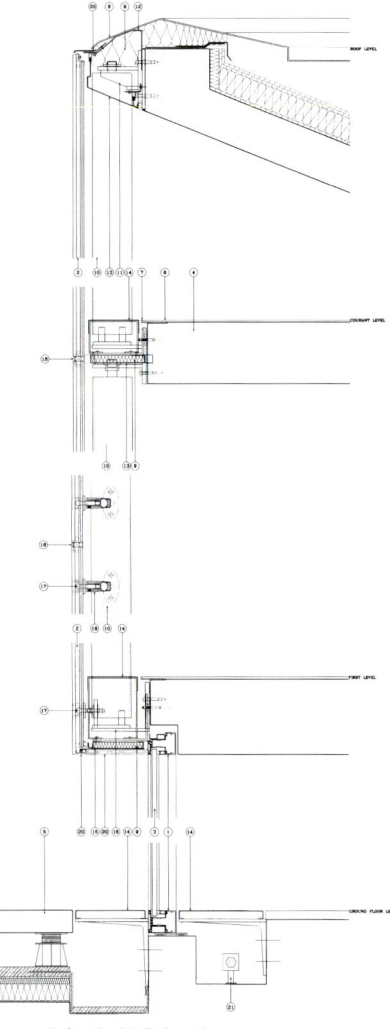

north facade detailed section

1. polished aluminium profile
2. printed tempered clear glass^{THK10mm}, air space^{THK15mm}, laminated clear glass
3. laminated clear glass
4. steel reinforced concrete slab
5. precast concrete element
6. poly vinyl flooring
7. stainless steel angle
8. aluminium protection sheeting
9. insulation
10. elliptic section polish stainless steel post
11. anchor plate
12. galvanised steel sheet^{THK30, 10mm}
13. polished aluminium sheet^{THK20, 10mm}
14. stainless steel perforated panel
15. galvanised steel panel^{THK20, 10mm}
16. galvanised steel plate^{THK 10mm}
17. articulated glass-anchoring thru bolt
18. black silicone water proof joint
19. cast-stainless steel anchor
20. extruded synthetic rubber joint
21. heating tube

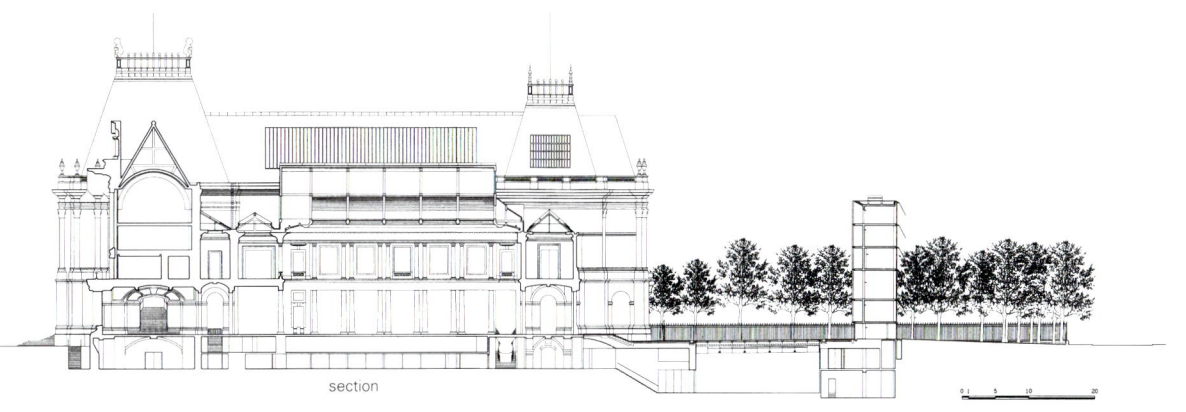
section

Project architect in charge: Pierre Cantacuzène
Project architect for facades and muséography: Sophie N'Guyen
Structural Engineer: Khephren Ingénierie
Facade consultant: Y.R.M. Anthony Hunt Associates
M.E.P. Engineer: Alto Ingénierie
Specification consultant: Atec
Safety consultant: Cabinet Casso & Cie
Lighting consultant: L'Observatoire 1 - Georges Berne
Signage designer: Visual Design - Jean Widmer
Landscape consultant: Louis Benech
Client: Town of Lille
Net surfaces: Renovation and extension under existing - 17.000m²
New addition and underground expansion - 11.000m²
Cost: Base building - 150 millions francs taxes not included
museography - 30 millions francs taxes not included
Program: Permanent exhibits - Sculpture, Paintings, Graphic arts, 17th centuary military models, Ceramics
Temporary exhibits
Services - Reception, Bookshop, Café, Teaching department, auditorium, Library, Curators offices
Photograph: Georges Fessy, Hervé Abbadie

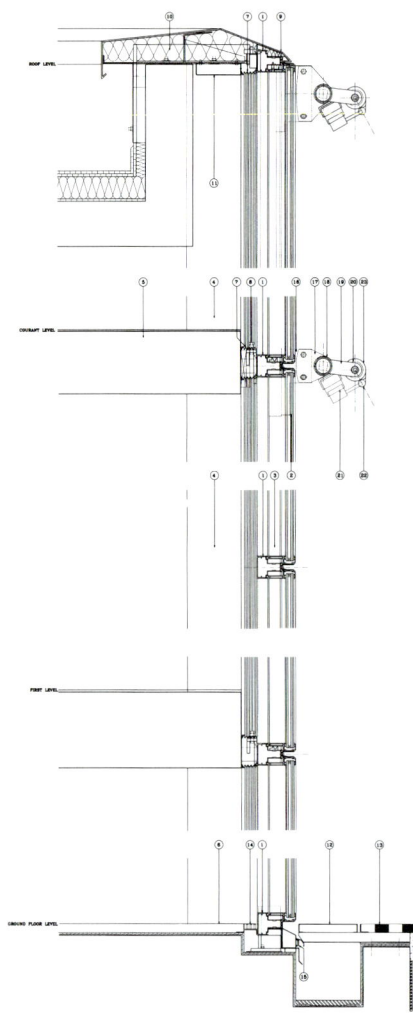

south facade detailed section

1. polished aluminium profile
2. matrix glass THK4mm, laminated planitherm THK8mm
3. polished aluminium mullion
4. steel reinforced concrete column
5. steel reinforced concrete slab
6. polished concrete flooring
7. extruded synthetic rubber joint
8. adjusting bolt
9. aluminium protection sheeting
10. insulation
11. galvanised steel sheet THK20, 10mm
12. stainless steel perforated panel
13. red color gravel and asphalt mix
14. aluminium grid
15. self opening air valve
16. stainless steel bracket THK8mm
17. polished stainless steel plate THK8mm
18. polished stainless steel supporting tube diameter 70mm, THK8mm
19. polished stainless steel tube supporting plate THK8mm
20. motorised tube
21. polished aluminium arm
22. stainless steel weight rod diameter 25mm
23. white plastified fabric

오늘날 가장 신기한 현상이자 우리에게 가장 흥미로운 현상은 바로 도시의 발전이다. 보다 정확히 표현하자면 완벽한 발전을 추구하려는 세계 곳곳 현대도시의 시도를 어떻게 지금처럼 회피할 수 있느냐가 바로 우리의 관심거리다. 도시를 계획하고, 구성하고, 조직화함으로써 완벽한 도시를 만들려고 애쓰면 애쓸수록 결과는 예상과 빗나가게 마련이다. 우리가 맡게 된 것은 도시의 추상, 분절된 도시다. 바로 눈앞에 놓여있는 자연을 실제로는 정의내리기가 어렵다는 사실 앞에서 놀라지 않을 수 없다.

전형적인 도시는 수직적으로 통제되고 조직되어 제어할 수 있는 실체로서 역시 수직적으로 통제되고 조직된 사회의 모습을 반영한다.

이러한 도시에서는 전체만이 의미가 있다.

고대 도시는 추상적이고 초월적이며 반감될 수 없는 절대적인 미의 개념에 따라 구성되었다. 우리가 받아들이고자 하는 것도 바로 그러한 도시개념이다. 그리고 현대도시들이 더 이상 우리가 표방하는 다양성과 복합성을 대변하는 데 유일한 것이 아니라는 점도 잘 알고 있다. 오늘날의 세계를 그러한 비전에 들어맞도록 하는 것도 이상적으로나 가능할 것이다.

사물에 제약을 가하면 어디선가 또 다른 문제가 예기치 못하게 다시 나타날 수 있다. 따라서 우리는 사물에는 내적인 가치가 있으며, 그러한 내적 가치 안에는 현실과 직접적으로 연결된 고유의 아름다움이 있다는 것을 인정하는 또 다른 접근방식을 선호한다. 미는 우주 만물의 본질에 깃들어 있으며 보는 이의 관점에 따라 상대적이라고 생각한다. 이러한 상대적 미학론은 무한한 가능성을 부여한다. 그런 관점에서 복원작업을 긍정적인 방향으로 고려할 수 있다.

우리의 복원 전략의 골자는 기존의 자질들을 보강하는 것이었다. 건물이 지니고 있는 잠재력을 끌어내어 경제적으로 작업을 한다는 것이다. 추상적 개념을 도입하려고 애쓰는 대신 생명력을 불어넣는다. 그리고 작업공간에 걸맞은 노력을 기울인다. 그러면 공간들을 변화시키고 통합하고 발전시키는 작업이 한결 수월해진다.

우리 작업방식에서 중시하는 경쾌함과 융통성은 용도와 해석에 있어서 무한한 가능성을 지닌 다양한 개념을 창출한다. 우리의 설계에 현실세계와 통합될 수 있는 무형의 요소가 내재돼 있기를 바란다.

릴 미술관은 훨씬 큰 규모의 미술관을 지으려던 당초의 계획을 축소하여 1892년에 건립된 것이다. 베랄드와 델마가 설계한 원안은 실제 공사된 건물의 두 배 크기로, 발미 가에 이르는 전체 부지를 아우르는 것이었다.

1930년대에 만들어진 중정을 아트리움으로 개조하면서 유리 천장을 만들었다. 그 아트리움이 건물의 구심점 역할을 하면서, 아트리움을 둘러싸고 미술품이 전시되었다. 그리고 2층의 회화전시관들로 통하는 초대형 계단을 증축한 후 남쪽 아케이드를 안쪽에서 폐쇄했다.

1970년대에는 행정동을 증축하면서 외부의 남쪽 아케이드를 폐쇄했다. 동시에 도청을 향하고 있는 북쪽 윙부분에 위치한 대형 위카르 전시관을 3개 층으로 분할하고 그 중 2개 층은 미술품창고로 활용했다.

공간이 협소하고 설비가 부족하다는 것도 새로운 건물을 신축하게 된 요인으로 작용했다. 이런 맥락에서 앵발리드 미술관의 릴리프 작품 컬렉션의 일부를 수용하기 위해 1980년대 말에 제안된 계획안은 범세계적 차원에서 미술관을 다시 돌아볼 수 있는 기회를 부여하였다.

1990년에 개최된 설계경기에서는 릴리프 작품의 새로운 전시공간 및 임시전시관·강당·도서관·어린이작업실 건립은 물론 행정동 개편을 그 내용으로 하였다. 아울러 기존 건물을 개조하여 회화, 중세미술품, 도자기류, 19세기 조각품 전시를 위한 새로운 컨셉의 제안이 요구되었다.

릴 미술관은 도심지의 노른자위 대지에 자리잡고 있다. 같은 시기에 속하는 두 건물은 도청을 향하고 있으며 서로 영향을 미치고 있다.

당초 계획했던 규모의 절반으로 건축되어 전반적으로 공간이 밀집돼 있기 때문에 후에는 다소 문제가 되었다. 메자닌 층이 있는 전시관들은 길이 방향으로 구획되었으며, 정원측 파사드는 1970년대에 건물을 증축하면서 가리워지게 되었다. 프로젝트의 기본 의의가 훼손된 것이다.

이 미술관은 부피, 빛과 그림자, 오늘날에는 상상할 수 없을 정도로 고급스러운 공간, 위풍당당한 분위기와 탁월한 내부 균형의 연속을 특징으로 하는 위엄있는 건축물이다. 이처럼 유서깊은 건물을 제대로 복원하지 못한 채 기존의 틀에 너무 얽매일 위험이

있었다. 그래서 우리는 미술관을 가능한 한 보존하는 대신 그 품격을 높이기로 하였다. 어떤 면에서 보면 전반적인 증축 프로젝트가 이런 기본지침을 바탕으로 하였다고도 볼 수 있을 것이다.

미술관의 탁월한 위치를 최대한 활용하여 살아있는 미술관이 되기 위해서는 무엇보다도 당대를 향해 열려 있는 미술관이어야 한다는 신조를 가지고 미술관을 도시 전체에 활짝 개방하기로 하였다. 따라서 시대에 걸맞지 않는 증축부분은 과감하게 없애는 반면, 원래의 건물구조를 복원하면서 깊이와 전망을 개선하였다.

1970년대에 증축된 부분과 아트리움을 복잡하게 만드는 중앙계단을 없애고 아케이드를 다시 개방하였다. 미술관을 외부지향적으로 바꾸어 1895년 베랄드와 델마가 맡았던 최초 프로젝트를 복원하였다.

우리는 이러한 복원과정을 통해 외부공간을 미술관 내부와 통합하였다. 조각 정원은 도청 대로변에 위치하는 하늘을 향해 열린 전시관이다. 전체 구성은 '평평한 판'이 수직 방향으로 여러 층 포개져 있는 형식을 취하고 있다.

그 첫 번째 판은 미술관의 인상주의적 이미지를 반사하는 투명유리로 구성된다. 그 표면 너머에는 붉은 배경에 금색의 모노크롬이 강렬한 대조를 이루고 있다. 이러한 금색과 붉은색의 대조는 미술관 표지판에도 활용되었다.

본 프로젝트에서는 옛 것과 새 것을 이어주는 가교로서 '반사'라는 개념을 사용하였다. 새롭게 재현된 이 릴리프의 이미지에서 19세기 건물의 릴리프가 반영되어 나타난다. 신축건물은 이러한 이미지를 확고히 하는 구조를 띠고 있다. 거울은 깊이를 강조함으로써 기존의 논리에 새로운 논리를 통합시킨다.

방문객들은 기존에 있던 두 개의 문으로 출입하게 되어 있다. 우리는 '열린 미술관'을 설계한다는 기본 취지에 부합하도록 사방에서 1층으로 자유롭게 들어와 아트리움을 지나 조각 정원과 까페 겸 식당으로 갈 수 있는 구조를 설정하였다.

미술품 관람은 아트리움을 구심점으로 3개 층에서 할 수 있다. 관람객이 집중되는 지점은 보다 힘있게 강조하면서 공간에 변화를 주어 관람객이 다양한 동선을 통해 미술품을 음미할 수 있도록 하였다.

각 공간마다 나름의 논리를 따르고 미술품들을 효과적으로 전시하기 위해 우리는 각기 다른 유형의 다양한 공간들을 만들었다.

위에서 조명되는 대형 전시관들이 있는 2층은 계속 회화진용관으로 사용하도록 하였다. 벽의 붉은색은 미술품과 웅장한 공간을 잘 조화시킨다.

석회암으로 만든 1층의 동쪽 및 서쪽 전시관에는 조각품과 도자기를 전시한다.

중세와 르네상스 미술품들은 아치형 천장의 지하전시관에 전시되는데, 이 공간은 높은 벽돌 천장을 깨끗이 닦고, 미술품을 전시하는 데 필요한 최소한의 부위에만 손을 댔을 뿐 거의 원형 그대로 보존하였다.

그럼으로써 전시관의 공허함과 리드미컬한 벽돌아치는 그대로 보존하기로 하였다. 종교화 위주의 이곳 전시품들은 커다란 오크목재 받침대 위에 전시되는데, 변화하는 표면을 창조하기 위해 받침대 상부에 색채를 가미하였다.

군대전시관 공간은 아트리움 아래를 굴착하여 마련하였다. 따라서 전시품의 규모에 걸맞는 자유분방하고 커다란 공간을 확보할 수 있었다. 이 전시관의 벽과 천장, 바닥은 검은색으로 통일하였다. 조명을 전시품 주위에 정확하게 배치하여 전시품을 전반적으로 볼 수 있도록 하였다.

우리는 살아있는 미술관의 개념을 충실히 구현하고 있는 현대예술가들의 작품을 두 개의 핵심 구역에 전시하였다. 출입구 별관에 달려 있는 게타노 패스케의 커다란 조명들은 '현실의 일상을 표현하는 잡동사니들을 담은 두 개의 가방'처럼 원형공간들을 잠식한다. 실크스크린 기법으로 처리한 채색유리 안에 48개의 입방체를 표현한 굴리오 파올리니의 작품은 아트리움을 압도하며, 동시에 "이곳이나 다른 어떤 곳에서도 전시공간과 작업공간, 빛과 침묵의 향연을 사로잡는 것은, 다시 말해 미술관을 형성하는 것은 물체가 아니라 우리의 관점이요 표현방법이다"라는 그의 생각을 잘 보여주고 있다.

우리가 당면해 있는 요소와 공간들의 정수를 제대로 파악하면 그것들을 보다 폭넓게 해석할 수 있다. 그러면 과거와 현재가 공존하면서 자유로운 가운데 서로를 풍요롭게 할 수 있게 되는 것이다.

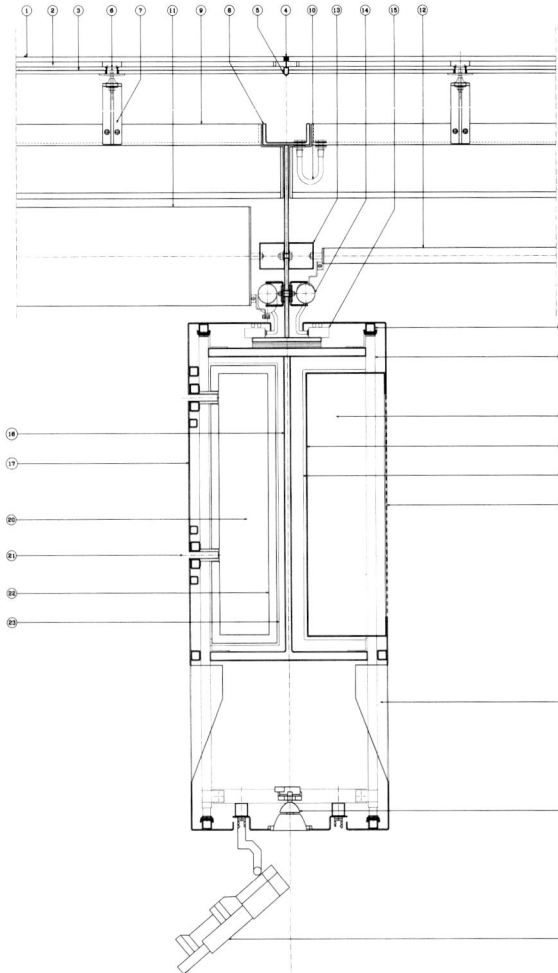

1. glass THK15mm float
2. air space THK15mm
3. tempered laminated glass
4. black silicone water proof joint
5. extruded synthetic rubber perimeter joint
6. articulation
7. cast steel anchor, bolted to steel chanel
8. structural steel chanel, welded to upper structure
9. structural steel chanel, welded to sub-structure
10. water drainage link
11. aluminium solar occultation blade - opened
12. aluminium solar occultation blade - closed
13. painted steel blade support
14. blade motor
15. electrical box
16. structural steel long span beam
17. painted steel casing
18. steel horizontal tube 30x30x2mm
19. steel vertical tube 30x50x2mm
20. HVAC duct
21. linear slot
22. fiber glass insulation THK25mm
23. fire proofing THK25mm
24. smoke duct
25. steel sheet THK3mm
26. fire proofing THK40mm
27. perforated steel casing
28. lighting fixture
29. downlight
30. spotlight

detailed cross section thru horizontal glass roof

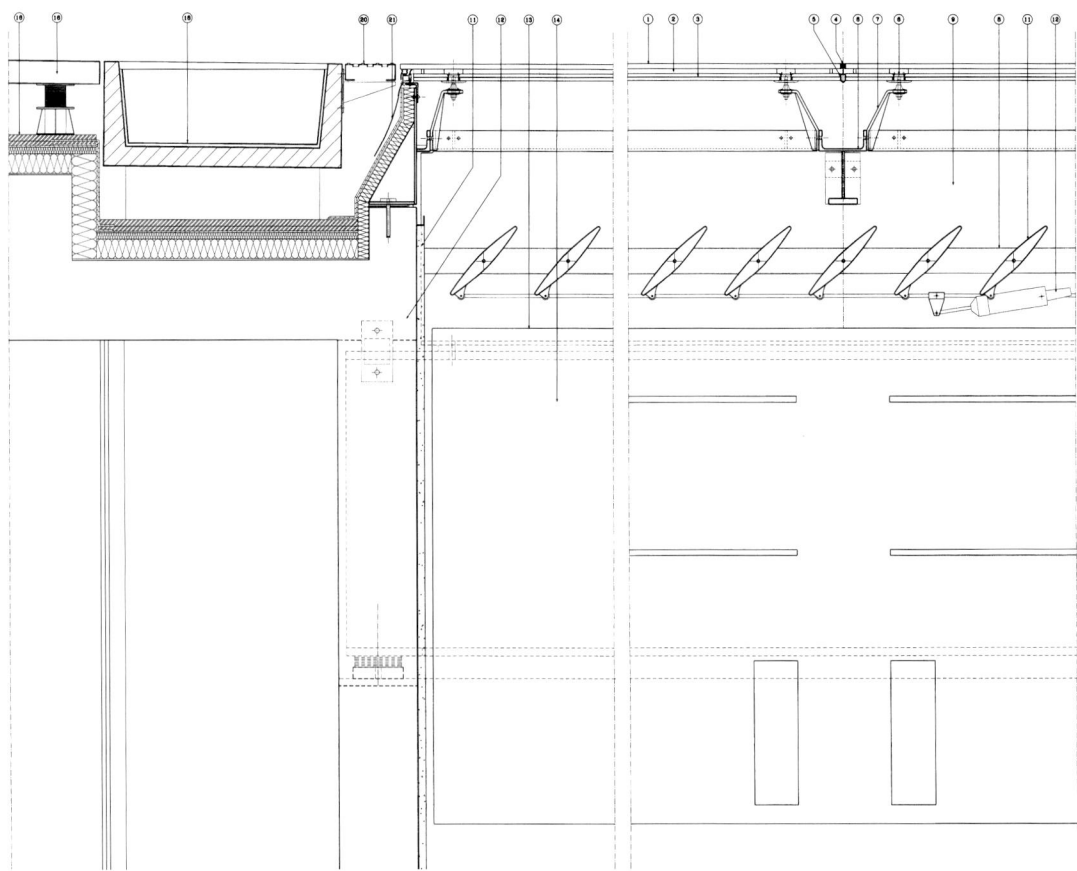

1. clear glass THK15mm float
2. air space THK15mm
3. tempered laminated glass
4. black silicone water proof joint
5. extruded synthetic rubber perimeter joint
6. articulation
7. cast steel anchor, bolted to steel chanel
8. structural steel chanel, welded to sub-structure
9. structural steel upper structure
10. painted steel blade support
11. aluminium solar occultation blade
12. blade motor
13. steel reinforced concrete structure
14. plaster covered, art hanging wall
15. painted steel casting
16. structural steel long span beam
17. precast concrete element
18. water proofing and insulation
19. precast concrete foutain
20. stainless steel perforated panel
21. synthetic rubber waterproofing sheet

detailed long section thru horizontal glass roof

New Wing of the Van Gogh Museum

반 고흐 미술관 신관

Kisho Kurokawa
쿠로카와 키쇼

The new wing was built in the open space adjacent to the main building of the museum, which was the last work of the Dutch Modernist architect Rietveld.

Considering the whole of the landscape, 75% of the building's area - excluding the main exhibition hall - was constructed underground as an effort to minimize the space it would have taken above the ground. The new wing connects to the main building through an underground passage.

Although Rietveld and Kurokawa share the Modernist idea of geometric abstraction, Kurokawa's new wing departs from Rietveld's linear style with curvilinear shapes and lines, employing a traditional Japanese idea of abstraction.

One characteristic expressing this idea is the sunken pool, situated between the new wing and the main building. Symbiosis between the new wing and main building is attained through the open intermediate space created by the pool.

The tilt of the elliptical roof and curve of the walls dislocate the center, underscoring the Japanese aesthetic of asymmetry.

Through highly abstract simple geometric shapes made complex, and further, through careful manipulation, the abstract symbolism of the new wing strikes a balance between the international and the local.

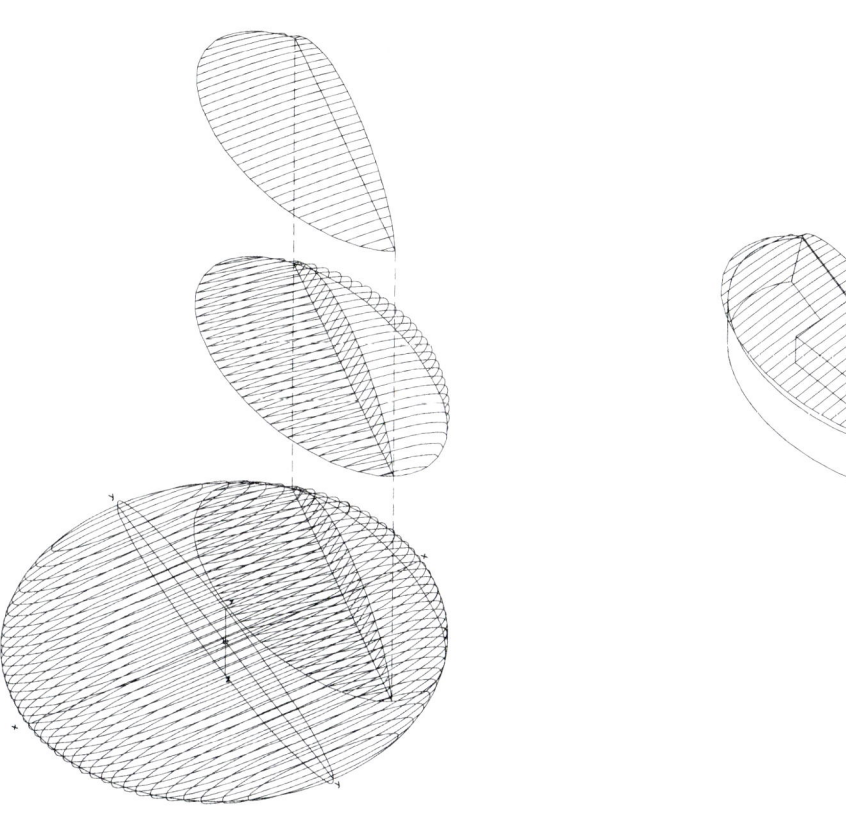

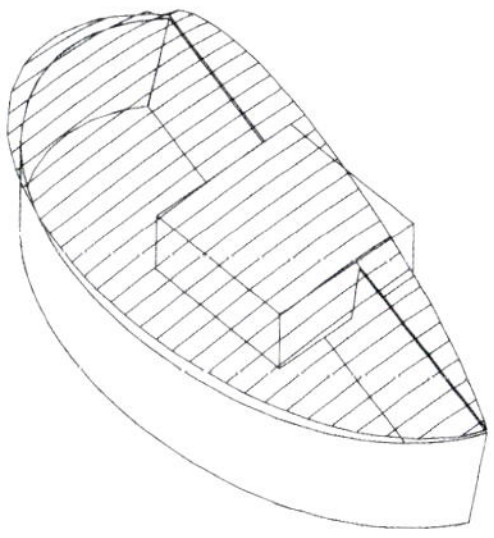

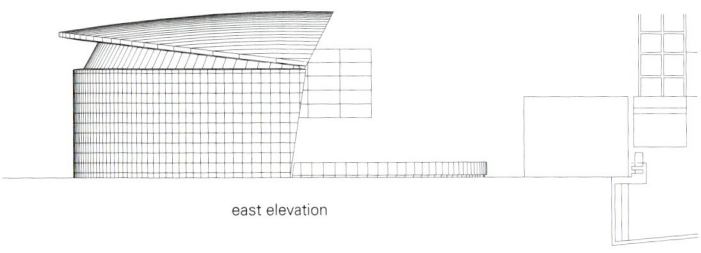

east elevation

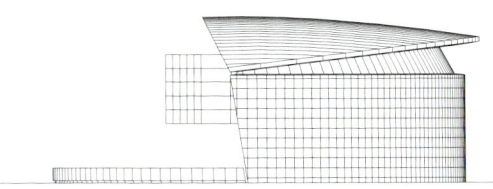

west elevation

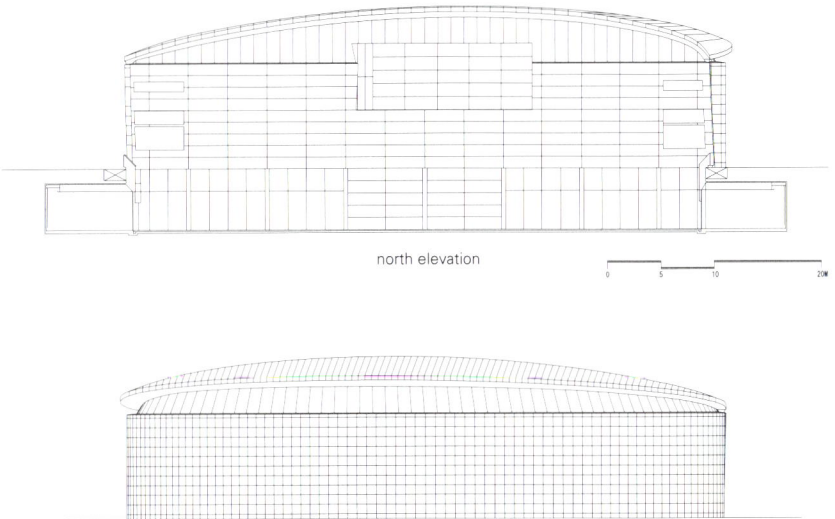

north elevation

south elevation

반 고흐 미술관 신관은 독일 모더니스트 건축가 리에트벨트의 미술관 본관에서 마주 보이는 광장에 만들어졌다.
전체경관을 고려하여, 지상의 건물을 최소화하기 위해 건물의 75%를 지하에 건설했다 (메인 전시관을 포함). 신관은 본관과 지하 통로로 연결된다.
비록 리에트벨트와 쿠로카와가 기하학적 추상개념을 바탕으로 건물을 설계하였지만 쿠로카와는 일본의 전통 추상개념을 도입함으로써 리에트벨트의 직선 스타일과 곡선을 이용한 형태에서 벗어나고 있다.
이러한 개념을 표현하는 한 형태로 본관과 신관 사이의 선큰인 물의 마당을 들 수 있다. 이 마당을 만듦으로써 열린공간을 통해 두 건물 사이의 공생이 이루어지고 있다. 기울어진 타원의 지붕과 벽의 곡선은 중심을 없애면서 일본의 비대칭 미를 강조한다.
신관은 수준 높고 추상적인 단순한 기하학적 형태가 혼합, 발전되는 능란한 작업을 통해 국제적, 지역적 균형을 상징하고 있다.

Location : Amsterdam, the Netherlands
Site area : 4,260m²
Bldg. area : 820m²
Total floor area : 5,000m²
Bldg. scale : One story below ground, two stories above ground
Structure : Reinforced concrete w/ Steel frame roof

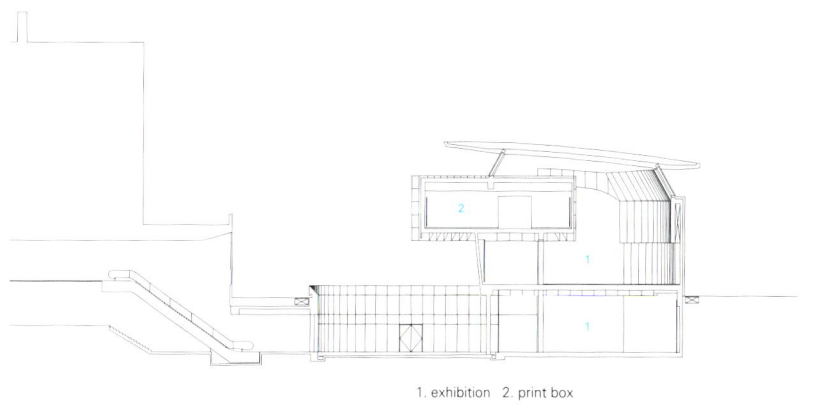

1. exhibition 2. print box

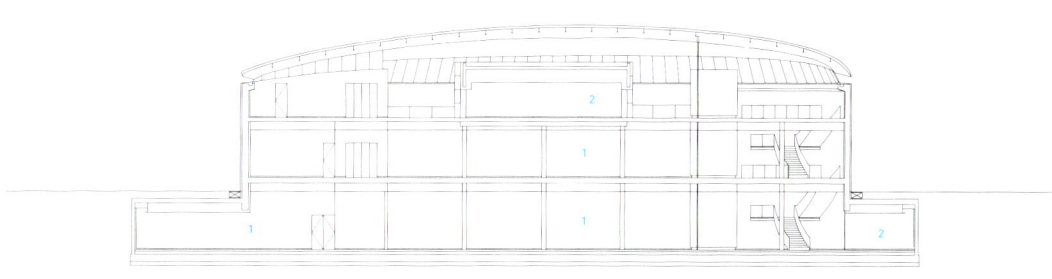

1. exhibition 2. print box

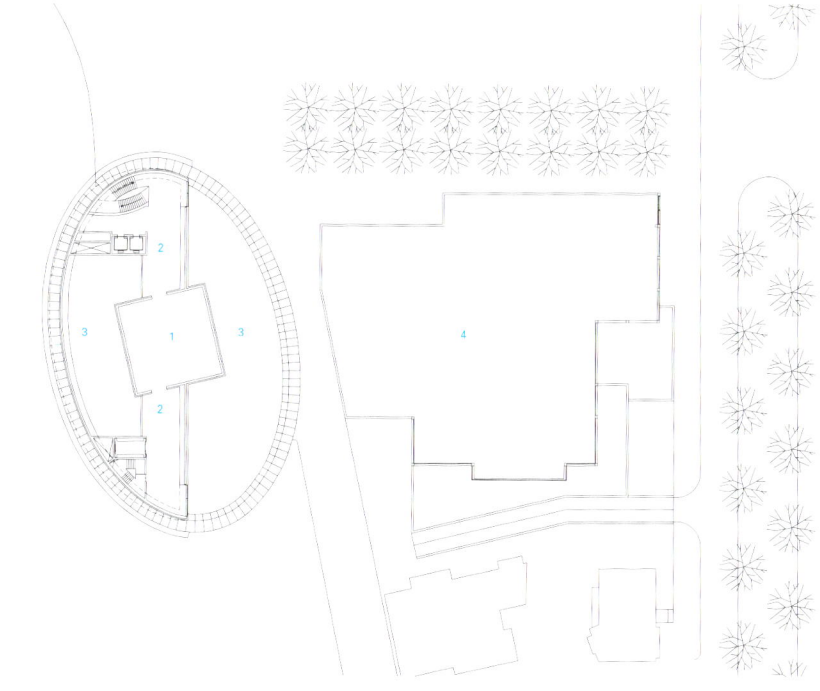

1. print exhibition   2. corridor   3. void   4. exhibition building

second floor

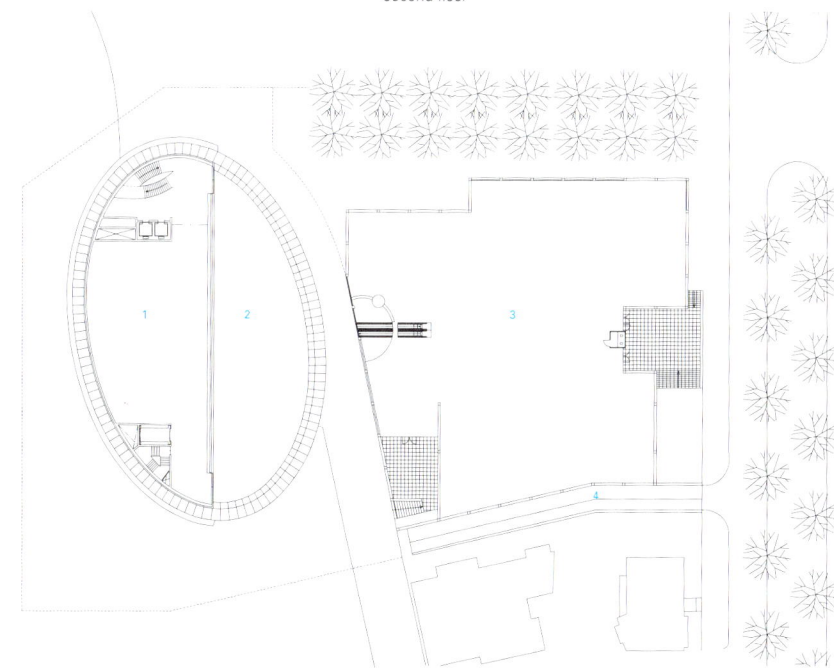

1. gallery   2. void   3. main entrance   4. entrance ramp

first floor

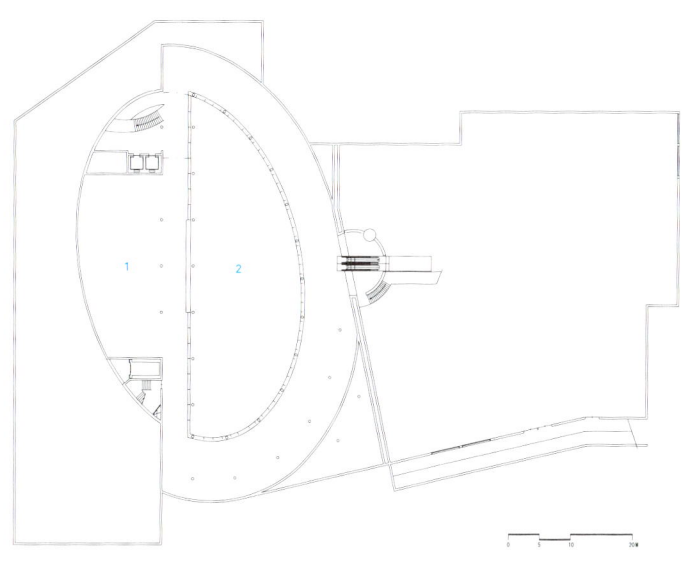

1. gallery   2. sunken water court

basement

Wolfsburg Art Museum
볼프스부르크 미술관

Schweger + Partner
슈베거 앤 파트너

Urbanistic space
At the southern end of Porschestraße, the south entrance to the city is being redefined with the new structure of the Art Museum in conjunction with the existing development, the Südkopfcenter and the extension of the town hall. The museum is conceived as a kind of urban loggia, its wide, dominating roof spanning various activities and marking the entrance to the city.

Placed in a field of tension between Hans Scharoun's theatre 1973 and Alvar Aalto's cultural centre 1962, the museum is to become a new focus with an identity of its own, creating new links without compromising the unique architecture of existing buildings.

The new museum stands on a huge 'stone slab', dividing it into comfortably sized piazzas connecting with the various paths and streets. A relationship between landscape and townscape is achieved with the slightly angular ramp at the end of Porschestrasse and the parallel underpass lined with a wall of water.

Organization and access
The museum roof spans a composition of simple volumes enclosing the exhibition hall like a forum. The entrance and lobby area with cloakrooms, a bookstore and a separate cafe creates a flexible, transparent zone for a variety of different activities, at the same time

visualizing the inside with a series of showrooms, and opens up onto the court of sculptures. The exhibition area is suited equally well to the gradually growing art collection and to changing high-quality exhibits. The concept also provides for several exhibitions to be staged concurrently, thus giving the curator a wide range of opportunities for enhancing the individual stagings by creating spatial connections and fascinating circulation paths.

The various possibilities of lighting design, from 'black box' spaces to zenith and side lighting, opens up a wide spectrum for presenting exhibits. Access to the museum is provided through an entrance hall in the shape of a rotunda facing Porschestraße. From this easily surveyed area, visitors follow several different stairways to the cafe to remain open to the public even after museum hours. The design studio, also accessible from the side facing the piazza, consists of a subdivided open space with additional gallery. The Art foundation building is organized onto two storeys with an open staircase.

The court of sculptures, providing spaces for experimental exhibitions and designed to be covered with a temporay roof, is closed off to the south by the workshops which will be used for in-house services such as restoration, packaging and storage and feature a number of working platforms. Air conditioning is provided from the central plant in the basement by way of horizontal distribution on the mezzanine floor, a crawlspace, and accessible ductwork surrounding the exhibition hall on three sides. Zones of different air quality will be established, ensuring a strict separation of foyer and exhibition areas.

The Art Museum will be protected by a comprehensive 'external security system' with additional motion detectors installed in strategic locations. The alarm system is controlled from the security control centre in conjunction with the surveying of the delivery gate and the personnel entrance.

Structure

The museum is based on a geometric design corresponding with the simple organization of the underground car-park.

The primary frame consists of a 8.10m x 8.10m grid, subdivided into 1.35m bays corresponding with the articulation of the facade and the roof.

The roof structure, a grate-like steel truss, is supported by a system of composite columns of steel and reinforced concrete spanning 16.20m on the inside and 24.30m on the outside of the regular columns bays. The cores, the stiffening shear walls and the floor slabs are designed as a reinforced concrete structure, while the suspended gallery consists of a grid of steel girders. The closed facades are clad with corrugated aluminum panels attached to a suitable substructure. The transparent parts of the facades consist of a post-and-beam glazing system in combination with glass sheets enameled on the back and mounted independently of the structural elements. The facade outside of the cabinet areas are articulate by adjustable aluminum sunscreen louvres mounted horizontally and vertically. The complex multi-level construction of the large museum roof is composed of a number of layers with different functions. The outer membrane consists of sandwich-type double glazing with a microgrid in between the glass sheets. Coinciding with the trusses, the walk-on gutters serve to facilitate roof maintance. Underneath the level of main trusses which provides passages in certain areas for installation purposes, movable scaffolds, subdivided into five sections, are mounted for main-tenance work. They provide easy access to the luminaires of the artificial lighting system as well as to the shielding grid mounted beneath, consisting of a system of reflecting aluminium lamellae for redirecting incoming daylight.

Throughout the museum, the mounting walls will receive interchangeable surfaces in the form of painted particleboard planks attached to a substructure with hidden wiring. The system of adjustable partitions will have the same surface quality.

Suspended ceilings for the exhibition areas, the cabinets and the gallery facilitate the installation of electrical wiring and luminaires. Catwalks, bridges and railings with various infils are designed as light-weight steel structures.

Daylighting and artificial light

In keeping with the geographic location of the building and the range of possible angles of incidence for direct sunlight in diurnal and annual rythm, a new

type of sunscreen grid is used.

Integrated into the multilayer glazing, this micro-grid consisting of plastic surfaces vapor-coated with pure aluminium reflects the angular portion of direct sunlight into the atmosphere. All other portions of the radion are transmitted to the interior through 'miniature lightwells'.

Based on theoretical findings of luminance studies, the shielding grid was designed to minimize the glare of the otherwise bright glazing.

At the same time, the grid serves to direct the incident light directly onto the objects on exhibit. The vertical illuminance is increased. The different width of the horizontal and vertical louvres are expressly designed to offset the symmetry of the ceiling, particulary along the edges and in the gallery area.

The artificial lighting systems are divided into two different milieus, serving both as permanently supplementary artificial lighting and as straight artificial lighting during the evening and at night, respectively.

The supplementary artificial lighting system utilizes secondary reflector luminaires equipped alternately with 55 W Dulux and 70 W HQI-TS lamps.

The light radiates upwards from the luminaires mounted above the support structure onto the vaulted secondary reflector and then shines downward where it is reflected by the shielding grid to illumine the interior.

Double focus luminaires, so-called nautilus spots are suspended between the lamellae of the lithting grid.

Supplementing the evening mood, the warm white 75 W halogen lamps serve to enhance the lighting milieu at night. The threshould angles of the spots coincide exactly with the edges of the shielding lamellae so as to prevent any unwanted brightening of the grid.

A zonal lighting concept for highlighting sculptures or pictures has been design-ed, consisting of a series of individual reflectors and spots. Parallel spots (1200 W and 2000 W respectively), mounted on the gallery, beam their light through the ceiling structure onto special reflectors which are individually controlled to provide zonal illumination of designated wall or floor areas.

All components of the artificial lighting system can be operated separately for the individual ceiling areas. The presentation of works of art is enhanced by a sophisticated lighting control system.

west elevation

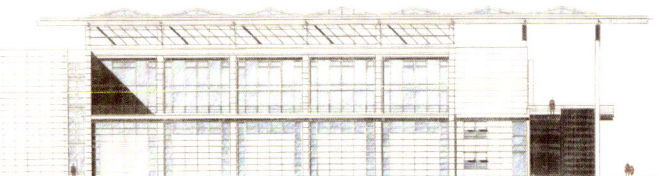
east elevation

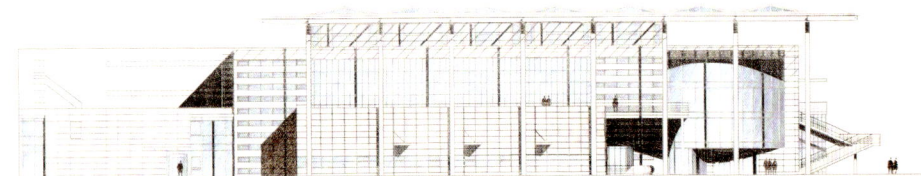
north elevation

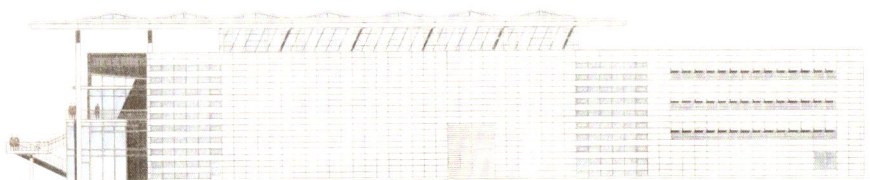
south elevation

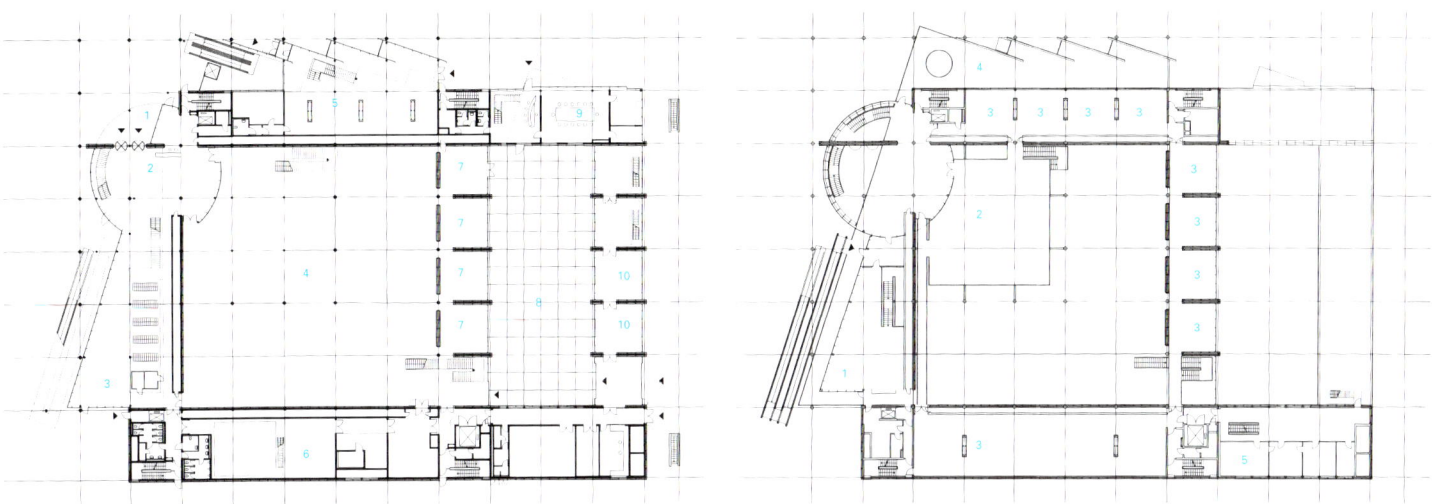

1. entrance 2. foyer 3. book shelf 4. hall 5. design studio 6. magazine
7. cabinet 8. sculpture court 9. conference 10. working room

ground floor

1. cafe 2. gallery 3. cabinet 4. terrace 5. administration

second floor

도시공간

포르쉐 가의 남단에 위치한 볼프스부르크 시의 남쪽 관문은 '주코프트센터와 시청사 증축'이라는 기존의 개발 프로젝트와 관련하여 공사가 이루어진 미술관의 새 구조로 인해 새로워졌다. 도시를 향해 열린 주랑의 개념으로 구상된 신축 미술관의 웅장한 지붕은 다양한 활동공간을 이어주는 동시에 도시의 관문이 된다. 1973년 한스 샤로운이 설계한 극장과 1962년 알바 알토가 설계한 문화센터 사이에 자리잡고 있는 볼프스부르크 미술관은 기존 건물들의 독특한 건축양식과 절충하기보다는 새로운 연결고리를 창출함으로써 나름대로의 정체성을 지닌 새로운 구심점으로 부상할 것이다. 볼프스부르크 미술관을 떠받치고 있는 거대한 석판은 여러 갈래의 도로 및 골목길과 이어져 있는 적당한 규모의 회랑들로 미술관을 분할하는 역할을 한다. 포르쉐 가가 끝나는 지점과 물이 흐르는 벽이 이어져 있는 평행한 지하도를 연결시키는 완만한 경사로를 통해 도시와 그 주변풍경이 자연스럽게 어우러진다.

구성 및 진입

볼프스부르크 미술관 지붕은 공공광장과 같은 주전시관홀을 둘러싸고 있는 단순한 볼륨들을 덮고 있다. 휴대품 보관소, 서점, 별도의 까페가 있는 출입구 및 로비구역은 다양한 활동을 위한 융통성 있는 투명한 공간을 제공한다. 동시에 일련의 전시실이 이어져 있는 내부를 시각화하는 한편, 조각품들을 전시하는 정원과 통한다. 전시 공간은 향후 소장품이 늘어나거나 전시품들을 교체하는 등의 변화를 충분히 수용할 수 있도록 구상됐다. 이 같은 탄력적인 개념 덕분에 여러 전시회의 동시 개최가 가능하다. 결과적으로 이것은 공간적 연계와 흥미로운 순환통로를 창출함으로써 큐레이터에게 개별적인 전시회를 장려할 수 있는 폭넓은 기회를 가질 수 있게 한다.

검은 박스형 공간에서 천정과 측면 조명에 이르는 상이한 조명설계는 전시품들을 다양하게 선보일 수 있도록 한다. 미술관으로 진입하려면 포르쉐 가를 향한 둥근 천장의 출입구출을 지나야 한다. 한눈에 들어오는 이 공간을 지나면 여러 개의 계단이 나오는데, 이것은 미술관 폐관 후에도 영업이 계속되는 까페로 이어져 있다. 회랑을 향하고 있는 측면에서도 출입할 수 있는 디지인스튜디오는 갤러리가 있는 열린 공간들로 나누어진다. 미술재단건물은 열린 계단이 있는 2개 층으로 구성되어 있다.

실험적 전시공간을 제공하며 임시가설지붕이 설치될 조각정원의 남쪽은 워크숍공간에 의해 막혀 있다. 이 워크숍공간들은 작품 복원, 포장, 보관 등의 자체 서비스를 위한 곳으로서 여러 개의 작업대를 갖추고 있다. 지하에 있는 중앙냉난방장치로부터 전시실의 삼면을 둘러싸고 있는 덕트, 바닥의 배관, 메자닌층의 수평배전구조를 통해 냉난방 및 환기가 조절된다. 따라서 각 구역의 공기를 달리 조절함으로써 로비와 전시구역을 엄격히 구분한다.

철저한 보안을 위해 필요한 곳에 동작 감지기를 설치하는 것은 물론, 종합적인 외부 보안시스템을 설치했다. 중앙통제실에서는 배달구역과 직원출입구를 모니터하는 것은 물론 경보장치도 통제하게 된다.

구조

볼프스부르크 미술관은 지하주차장의 단순한 구조에 부합하는 기하학적 설계에 토대를 두고 있다. 미술관의 중심구조는 가로, 세로 8.1m의 정사각형으로, 파사드와 지붕의 분할과 일치하는 가로, 세로 1.35m의 공간들로 나누어져 있다. 격자형 강철트러스로 구성된 지붕을 안쪽 16.2m, 바깥쪽 24.3m의 혼합식 기둥들이 떠받치고 있다. 중심구조 및 보강벽과 슬래브 바닥판들은 철근콘크리트구조로 설계된 반면, 떠있는 갤

러리는 격자형 강철 대들보로 구성된다. 기초공사를 한 후 물결무늬의 알루미늄패널들을 마감재로 사용함으로써 불투명 파사드를 만들어낸다. 이에 반해 뒤쪽에 에나멜을 입힌 판유리와 잘 어우러지는 유리기둥 및 들보는 투명한 파사드를 구성한다.

작은 진열실 구역의 바깥에 있는 파사드는 개폐 정도를 조절할 수 있는 알루미늄 버티칼 블라인드로 구성되어 뚜렷이 구분된다. 커다란 지붕의 복합구조는 상이한 기능을 지닌 여러 겹의 층으로 구성된다. 바깥쪽 구조는 초박형 격자구조의 양면을 판유리로 감싸는 샌드위치 형이다. 트러스와 같은 격자형 홈통은 지붕 관리를 한층 용이하게 해준다. 특정 구역의 설비를 보수하는 경우 통로의 역할까지 하는 주요 트러스 아래쪽에 있는 이동 가능한 골격들이 다섯 부분으로 분할되며 보수관리작업을 위해 설치된다. 이들 골격으로 인해 조명장치는 물론 유입되는 햇빛의 각도를 바꾸기 위해 그 아래쪽에 설치한 알루미늄 반사판들에도 쉽게 접근할 수 있다.

전선이 설치된 하부구조 위에 교체 가능한 파티클보드를 덧붙인 형태가 미술관 벽의 마감재로 쓰이게 된다. 이동식 칸막이 역시 동일한 마감재로 처리되었다.

전시구역, 작은 전시실, 갤러리의 매달린 천장은 전기 및 조명장치의 배선을 용이하게 한다. 좁은 통로, 다리, 난간 등은 경량의 강철 구조로 설계되었다.

자연채광 및 인공조명

건물의 위치와 하루 또는 연간 직사광선의 투사범위를 고려해 새로운 형태의 차광그리드를 사용했다. 여러 겹의 유리구조에 통합된 이 격자구조는 순도 100%의 알루미늄을 입힌 플라스틱 표면으로 구성되며, 직사광선의 일부는 대기 중으로 반사되고 나머지는 그 틈새를 통해 건물 내부로 흡수된다.

휘도 조사의 이론적 결과를 토대로 빛을 선별적으로 반사하는 격자구조는 유리의 번쩍임을 최소화하기 위해 고안되었다. 아울러 격자구조는 유입되는 빛을 전시품들에 직접 비춰주는 역할도 한다. 이에 따라 수직의 조도가 증대된다. 버티칼 블라인드의 폭을 달리한 것은 갤러리의 가장자리를 따라서 있는 천장의 균형을 상쇄하기 위한 의도를 담고 있다.

인공조명장치는 자연채광을 보완하는 조명과 저녁과 밤에 사용하기 위한 조명으로 나누어진다. 보완적 인공조명장치로 55와트 둘룩스와 70와트 HQI-TS 램프가 교대로 설치된 반사조명이 사용되었다.

반사조명장치에서 나오는 간접조명은 간접반사체를 향해 위로 비춰졌다가 격자구조판에 의해 아래쪽으로 반사되면서 내부를 밝혀준다.

빛을 반사하는 격자구조판들 사이에는 두 개의 초점을 지닌 집중조명장치가 매달려 있다. 따뜻한 느낌의 75W 백색 할로겐램프는 저녁과 밤의 안온한 분위기를 한층 돋워준다. 불필요한 빛을 차단하기 위해 빛을 반사하는 격자구조판의 가장자리와 빛의 초점이 형성되는 지점의 각도를 정확하게 일치시켰다.

조각 및 회화 전시품들을 강조할 목적으로 일련의 개별적인 반사조명과 집중조명을 적절히 사용하는 구역별 조명개념도 동원됐다. 1200W와 2000W의 평행집중조명장치를 갤러리에 설치, 천장구조를 통해 특수 반사조명으로 빛을 비추도록 한다. 그리고 특수 반사조명장치들은 특정한 벽이나 바닥들을 선별적으로 비추기 위해 개별적으로 제어되도록 했다.

인공조명장치의 모든 부분은 각각의 천장 구역에서 개별적으로 운용될 수 있다. 이는 고도의 조명제어체계를 통해 미술품을 보다 효과적으로 전시한다는 취지에서다.

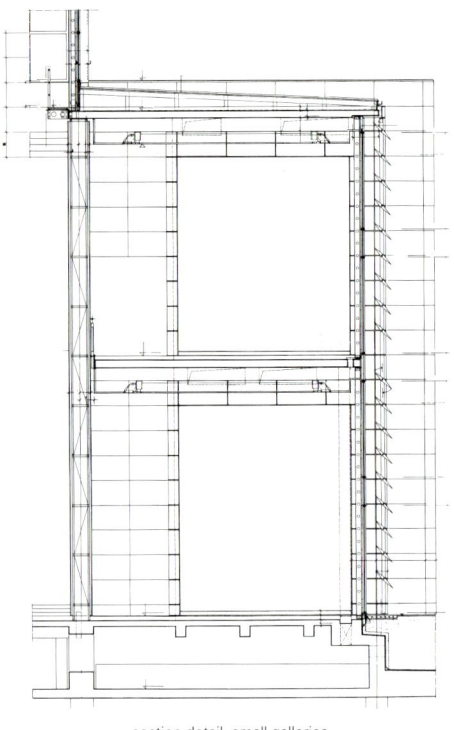
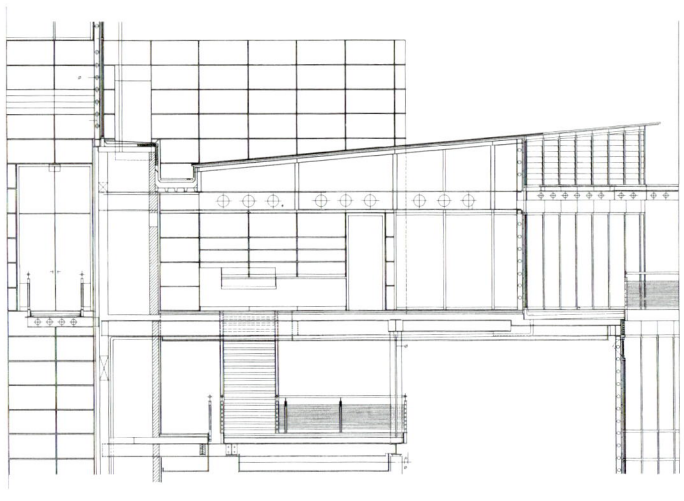

section detail, small galleries

section detail, gallery terrace

Project management: VW Wohnungs GmbH
Landscape architect: Prof. Gustav Lange
Structural design: bendorf+patner Ing.-Gesell.m.b.H.
Lighting design: Christian Bartenbach GmbH
Electrical planning: Ing.-Büro Paulus GmbH
Location: Wolfsburg, Germany

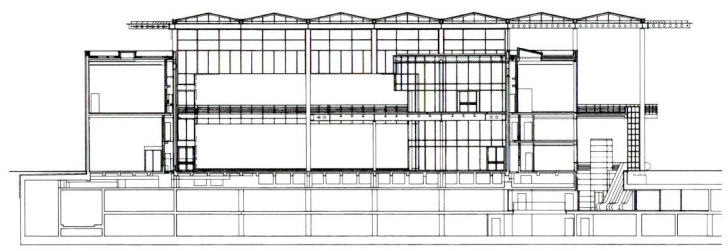

sections

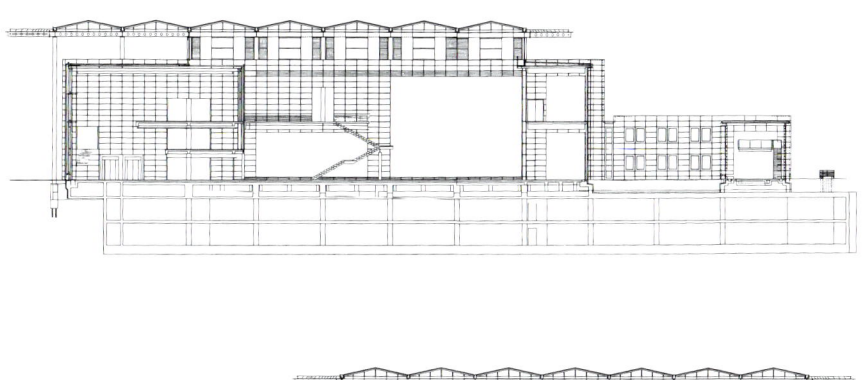
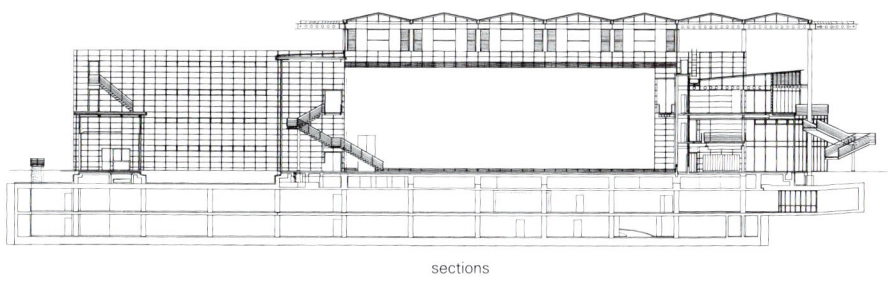
sections

Museum Sammlung Essl
삼믈룽에설 미술관

Heinz Tesar
하인쯔 테자

Anyone approaching this long, tall closed architectural structure on the unassuming access road or discovering it by chance from a passing train will have a difficulty in working out what it is for. This mysterious architecture has scarcely anything in common with modern building's familiar basic types for housing, offices or churches. It must be for some unusual purpose. And indeed the building does contain something that is most unexpected here, in the chaotically ill-built commercial area between the railway line and the meadows by the Danube : it is a museum of contemporary art, a cultural forum that has so far not existed in this form in Europe. This striking building contains what is probably the most comprehensive collection of recent art in Austria. It is private in every respect, that is to say it is run without any public subsidy ; and yet it has everything that one associates with a prestigious museum : storage space ; restoration workshops, its own museum team and also the necessary working capital.

Tesar has set standards above all with his many methods for providing light and his changing spatial configurations for the appearance rituals of contemporary art. His ingenious variation on the classical top-light lantern and his easy-to-use method for storing pictures should be part of the course at all institutes of higher education. But overall it is probably the free-and easy and yet highly intelligent linking and juxtaposition of the various museum functions that explain the charm of this compact spatial composition in Klosterneuburg.

Tesar's museum building creates something that is actually a contradiction : it puts itself entirely at the service of the works of art exhibited, and establishes itself as an architectural individual, as built work of art in its own right with precisely this special feature. Its enigmatic appearance gives some idea of the mystery that should always be part of the arts, but that has often been squandered in recent decades by some unduly agitated museum architecture.
written by Gottfries Knapp

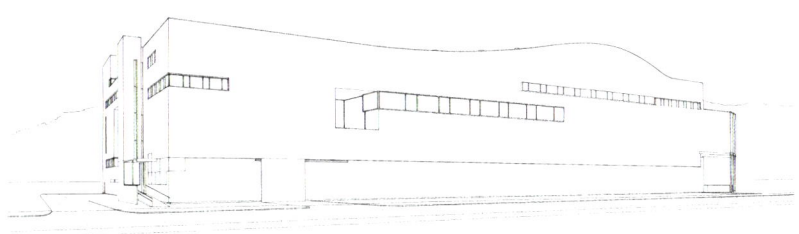

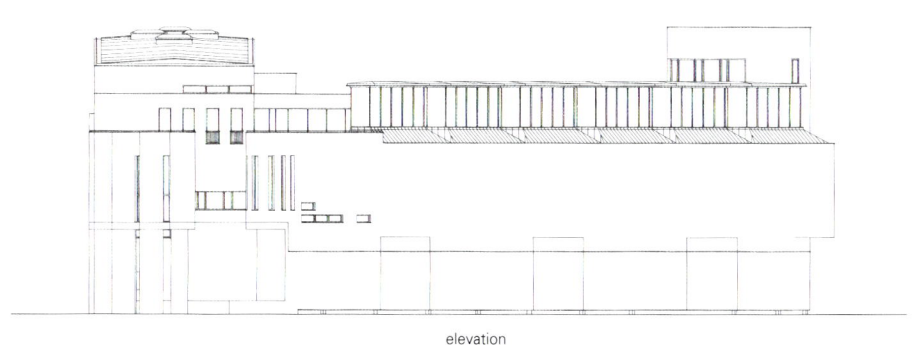

elevation

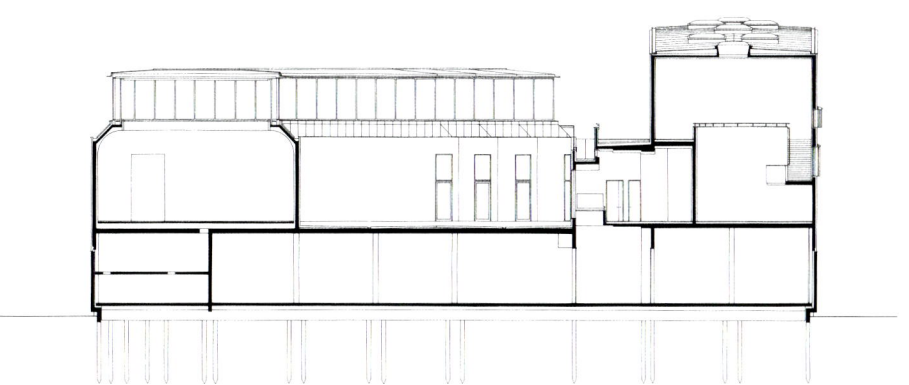

주위 환경에서 두드러져 보이지 않는 입구 도로에서 이 길고 높으며 폐쇄적인 건축물에 접근하거나, 기차를 타고 가다 우연히 이 건물을 발견하는 사람이라면 누구나 이것이 어떤 용도의 건물인지 알아내기 어려울 것이다. 신비로운 이 건물에서는 주택, 사무실, 교회 등 현대건물의 친숙한 기본 유형들과 공통점을 거의 찾을 수 없다. 이것은 뭔가 특별한 목적을 지녔음에 틀림없다. 실제로 건물에는 예상과 가장 동떨어진 것들이 담겨 있다. 철로와 다뉴브 강변의 초원 사이에 세워진 상업지구에 자리잡은 이 현대 미술관은 이제까지 이런 유형으로 유럽에 존재하지 않았던 문화포럼이다. 인상적인 이 미술관은 최근 오스트리아 미술계에서 가장 다양하고 많은 소장품을 담고 있다. 이곳은 보조금을 전혀 받지 않는 순수한 사설 미술관이다. 그러나 작품보관소, 복원작업실, 자체적인 미술관 인원과 필요한 운영자본에 이르기까지 유명한 미술관과 같이 모든 것들을 갖추고 있다.

테자는 무엇보다도 빛을 공급하는 여러 방법과 현대미술품의 외관 의식을 위한 공간배치의 변화를 우선시하며 기준을 정했다. 고전적인 상부조명의 풍부한 변화와 용이한 그림보관법은 모든 고등교육기관에서 가르치는 과목의 일부이다. 그러나 전체적으로 볼 때 이것은 얼핏 자유롭고 쉬워 보이지만 다양한 미술관 기능들이 고도의 지능으로 연결되고 늘어서서 클로스트노이부억에 꽉 짜인 공간구성의 묘미를 말해주고 있다.

이 미술관은 사실상 모순을 만들어 낸다. 미술관은 전적으로 전시된 미술작품을 위해 서있기도 하지만 특별한 용모를 지니고 세워진 예술작품으로서, 미술관 자체가 하나의 건축적 개체이다. 수수께끼 같은 외관은 모든 예술품의 일부가 되어야 하면서도 과도하게 동요된 미술관 건축에 의해 근래 십수 년간 종종 남발되었던 신비스러움을 던져 준다. 글 / 고프리드 크나프

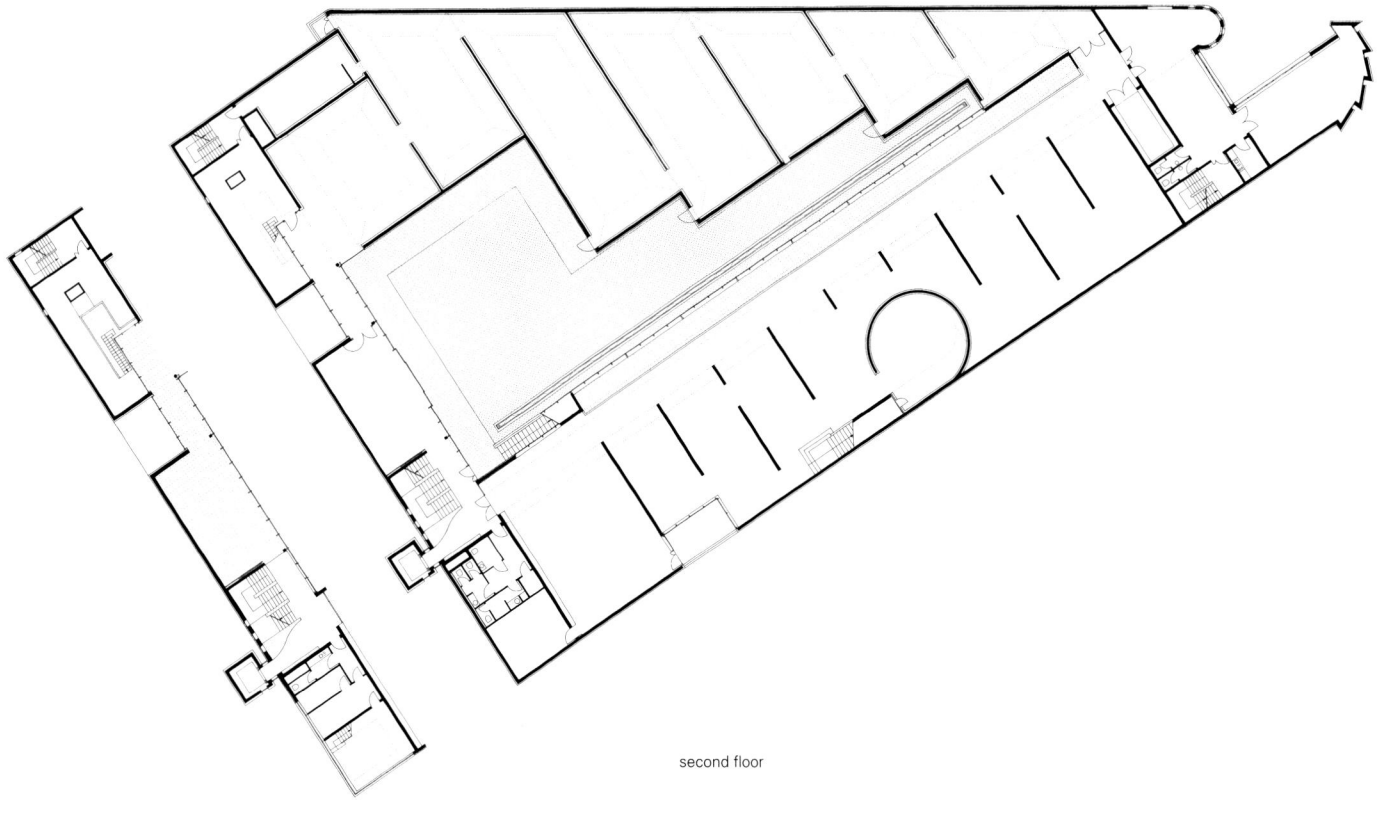

second floor

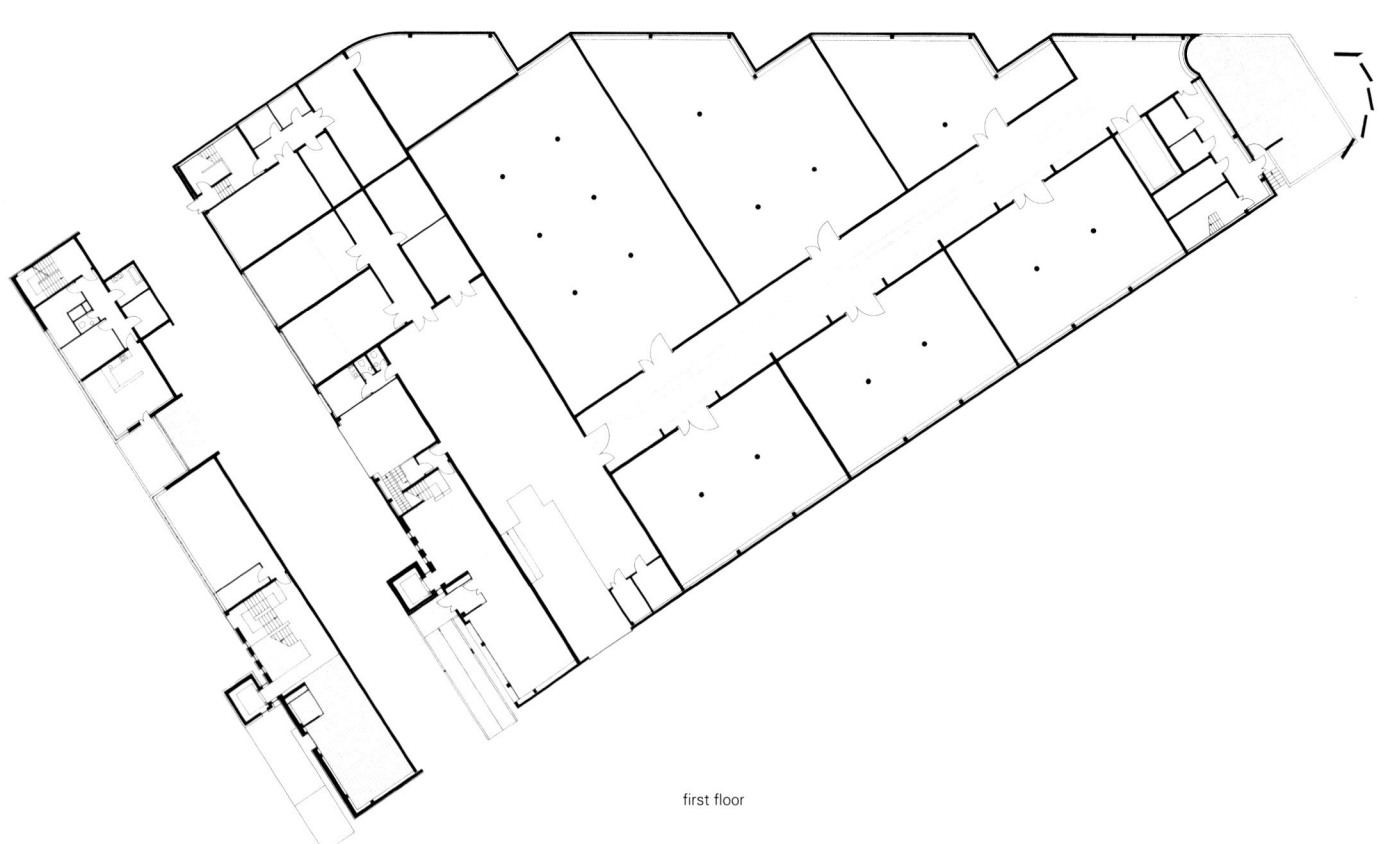

first floor

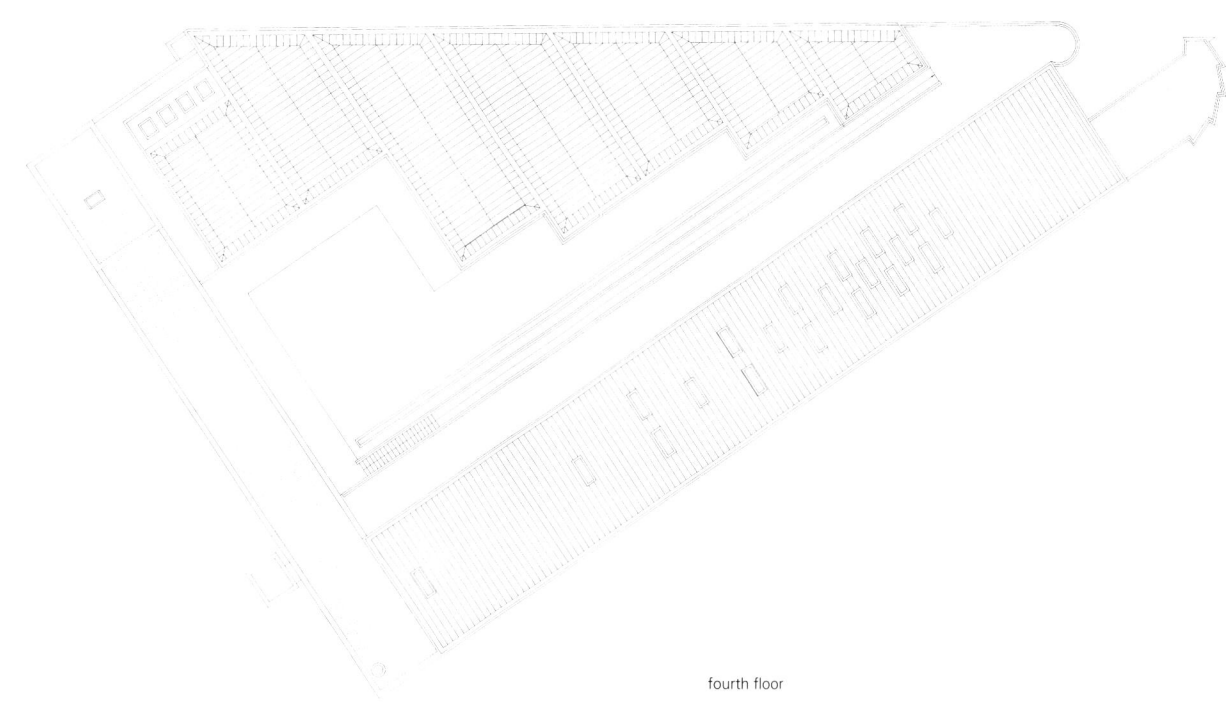
fourth floor

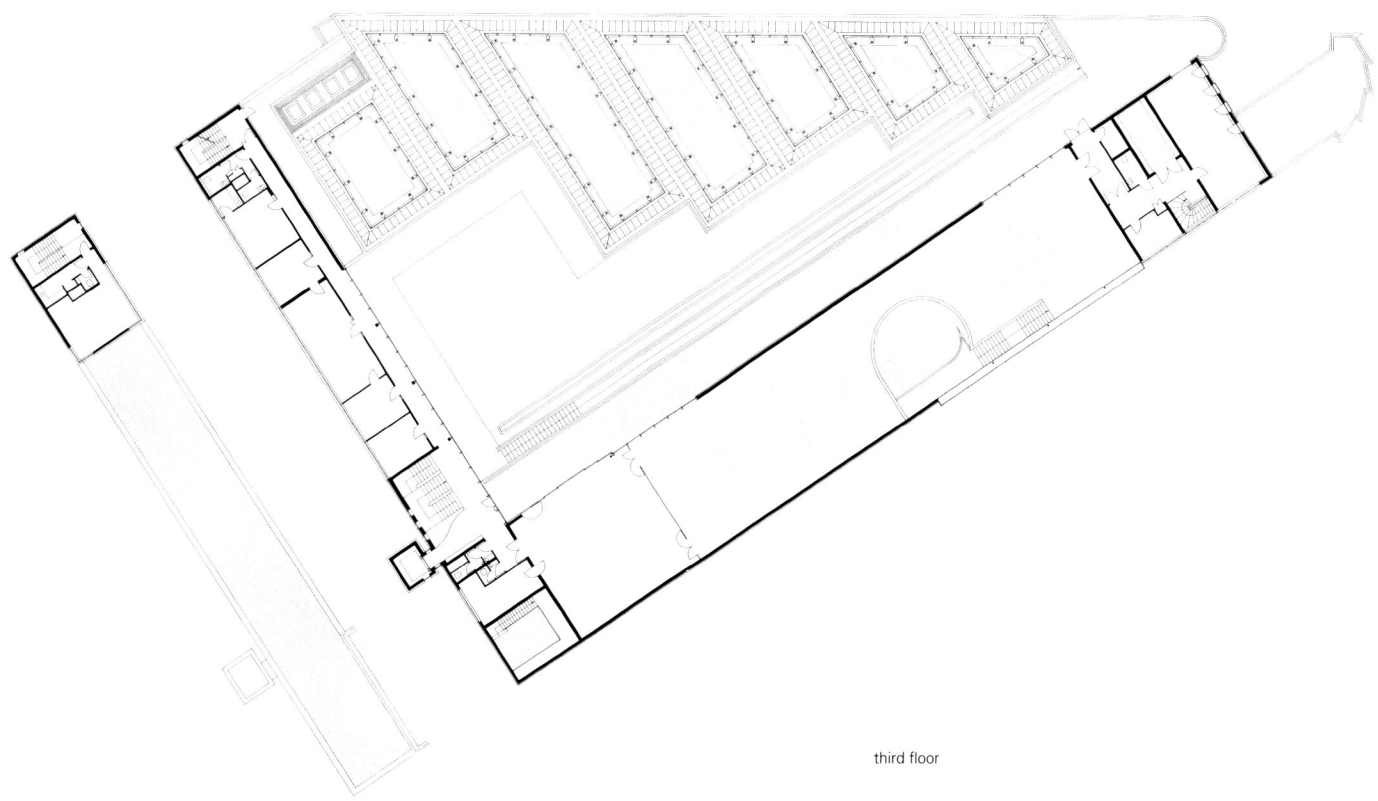
third floor

Photograph : Christian Richters - p.168, p.169, p.174top, p.175,
Margeritha Spiluttini - p.174bottom

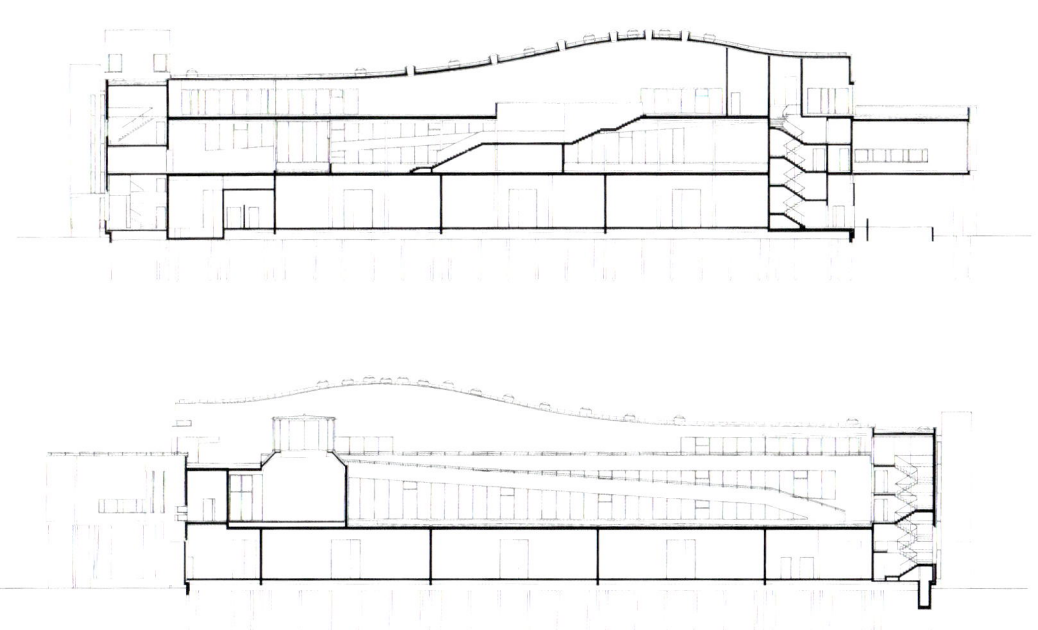

175

Australian Centre for Contemporary Art

호주 현대예술센터

Wood Marsh Architecture
우드 마쉬 아키텍춰

The Malthouse Development forms a precinct with the existing Malthouse Building which houses the Playbox Threatres. The Australian Centre for Contemporary Art, Chunky Move and Playbox make up the user group's for the new building. The ACCA has four gallery spaces off an entry foyer with service areas and office space. Chunky Move comprises two rehearsal studios and administration space and Playbox has a large set construction facility. A courtyard with amphitheatre is created at the centre of the overall complex with an outdoor exhibition space along the northern boundary.

This building is designed to make reference to its function...a sculpture in which to show art. It is also meant to support with optimism art practice and in a frugal sense, be a robust laboratory for experimentation. Those using the building would feel comfortably challenged. Openings in the external fabric are kept to a minimum to support installations, ephemeral and digitally projected work. There are also references in the building to past occupation of the site namely warehouses, foundries, etc. and the predominant shed vocabulary of steel frame and taut metal skin. The cladding is a single material to reinforce the form of the object.

몰트하우스 개발은 플레이박스 극장들이 있는 몰트하우스 빌딩이 위치한 문화예술단지다. 호주 현대예술센터 ACCA 와 무용단 청키무브, 플레이박스 엔터테인먼트 등이 신축 건물에 입주하게 된다. ACCA에서 각종 서비스 공간과 사무실, 출입구 로비를 지나면 네 개의 전시 공간이 있다. 청키무브는 두 개의 리허설 스튜디오와 행정실로 이루어져 있으며, 플레이박스에는 커다란 세트 제작실이 있다. 북쪽 경계선을 따라 야외 전시공간이 있는 전체 복합시설의 중앙에는 원형극장의 안마당이 있다.

ACCA 건물은 미술작품을 전시한다는 내적 기능을 외적 형태로 웅변하기 위해 조각품 같은 구조를 띠고 있다. 아울러 예술행위를 최적의 환경에서 지원하고 활발한 예술 실험의 장을 제공한다는 취지도 갖고 있어, 이용객들은 편안함 속에서 도전정신을 느끼게 될 것이다. 외관은 작품들을 보호하기 위해 창을 최소화했다. 창고, 주물공장, 격납고 같은 철골조와 단단한 금속 외관은 과거 이 건물의 용도를 암시하며, 단일 재료를 이용한 건물은 형태를 한층 부각시킨다.

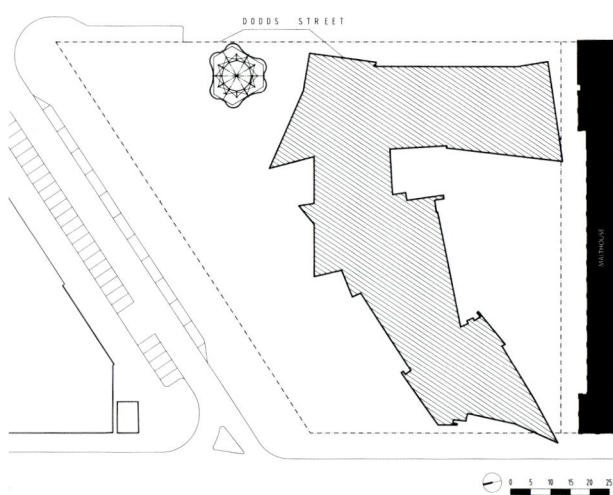

Bldg. area : 3,300m²
Construction : Concrete slab and piles, Steel frame, Corten cladding, Metal deck roof

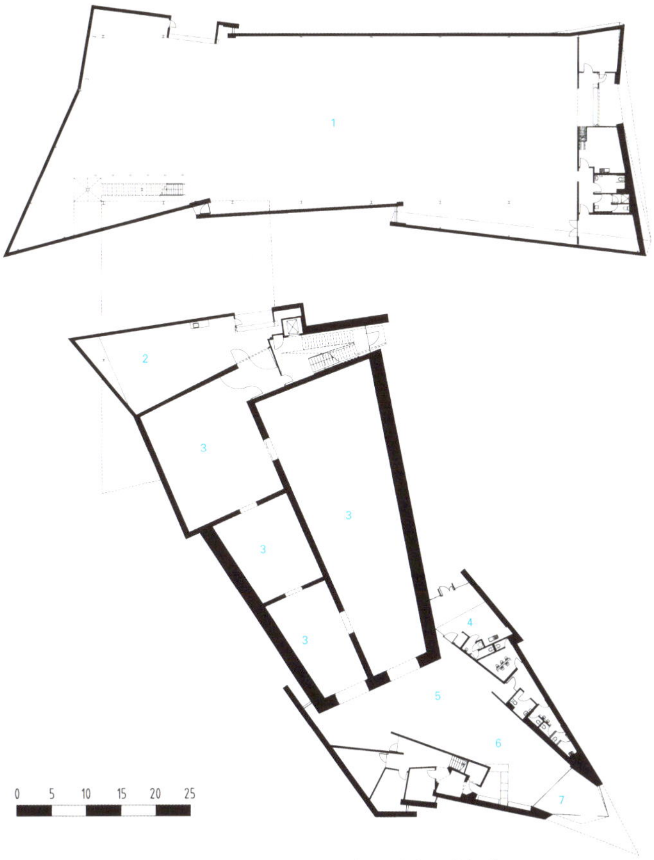

1. set construction workshop 2. loading
3. gallery 4. bar 5. foyer 6. reception 7. entry

first floor

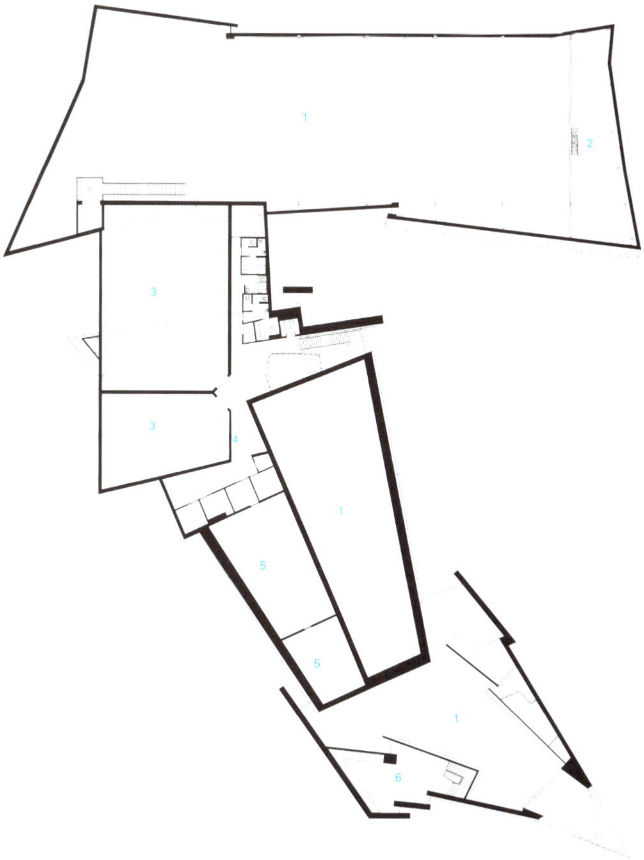

1. void 2. storage
3. studio 4. reception 5. plant room 6. office

second floor

Siida
사미 박물관과 북부 라플란드 방문객 센터

Juhani Pallasmaa
주하니 팔라스마

The Building
The museum is located next to a river at the edge of Inari village 300km north of the Polar circle. The two-storey building is placed on the southwest border of the open-air exhibition area of the Sámi Museum, which was established in 1959. The site of the building was formerly a gravel quarry and thus the building corrects the profile of the landscape. The main museum floor of the building, which is on the same level as the existing open-air museum, is entered along a gently sloping ramp leading up from the entrance lobby floor. At the point where the ramp turns, the visitor finds himself in the vicinity of the most impressive item of the open-air museum, an indigenous village courtyard. The basement floor contains the entrance lobby, an auditorium, a temporary exhibition space, the offices and guide facilities of the Visitor Center, and the museum's storerooms, workshops and utility rooms. The top floor houses the exhibition spaces proper-an introductory exhibition, touring exhibitions and a permanent exhibition on survival under extreme conditions-, a cafeteria for the public, a library and the Sámi Museum's offices. The route to the open-air museum leads past the ticket office and security desk through an outdoor ramp.

The external shape of the museum is articulated both horizontally and vertically to blend in with its small-scale landscape. The curving shapes of the roof follow the rolling countryside, and snow drifts during the long winter season. The protruding, curved eaves help to reduce the impression of height and give the building its humble character. A skylight runs the entire length of the building's midline, bringing light to the centre of the building. The building frame is of reinforced concrete and steel; the exterior walls are covered with varied wooden claddings.

The areas for the exhibitions are midnight blue, the other spaces are white. Mahogany-stained wood is used as a contrast with the brightness of the spaces. In the public areas the floors are of polished concrete dyed in the mix with a terracotta pigment. The lower surface of the reinforced concrete subfloor, cast with panel formwork, has been left visible.

Except for the lights in the cafeteria and technical fittings, all the light fixtures were designed specially for the museum. The indirect general lighting on the main storey uses the curved ceiling shapes as reflective surfaces.

The Exhibition
The contents of the exhibitions were under preparation for several years with a special committee of experts. The exhibition design progressed by the same architects in parallel with the building design. A full-scale mock-up section of the exhibition system was put together a year before the opening to test the many technical display solutions, such as the lighting.

The goal of the exhibition design has been to combine the scientific context with the experiential aspects of theatre or art exhibitions. The introductory exhibition provides a lead-in to the story of natural and cultural evolution in the north during 12000 years from the most ancient finds to the present day. The main exhibition contains nested presentations of Northern Lapland nature and of Sámi culture so that the overall spatial concept is one of cultural-ecological interaction. The shapes and strategies of life in the extreme harshness of the north are set out in an exhibition according to the seasonal cycle. Moving around the exhibition in the direction of the cycle thus gives a view of the annual rhythm of natural phenomena and the Sámi way of life, whereas moving from the outer circle inwards allows one to

1. Sámi museum and Northern Lapland visitor center 2. Sámi outdoor museum
3. storage building 4. custodian's lodging

read the linkages and interactions between nature, cultural heritage and the present day spirit of Sámi. A natural context for the Sámi way of life is built up with giant, backlit photos going around the outer walls of the exhibition room, showing the monthly variations of the year in nature, climate and light in northern Lapland. Photos, drawings, texts and videos are countersunk into these background images. Each month contains basic climatic information and a find-and-explore feature for children. The interface between the nature and culture exhibitions comprises a display case zone with subjects common to both domains ; looking from the outer circle, the exhibit shows a particular natural phenomenon, but when viewed from the raised central floor the same exhibit is linked with a cultural narrative. The central floor has backlit photographs sunk into it, on top of which there are installations depicting main features of the Sámi way of life. The exhibition space is dimly lit and the lighting focuses on the exhibits. The skylight strip in the ceiling divides the space into halves representing summer and winter.

The experience of the exhibition is, like an excursion into nature, a personal exploration with all its surprises. The photographs, drawings, objects and texts are augmented by a soundscape of the different seasons and natural occurrences arising in a freeform rhythm - some of it triggered by the movements of visitors - like the multisensual experience of moving around a natural environment.

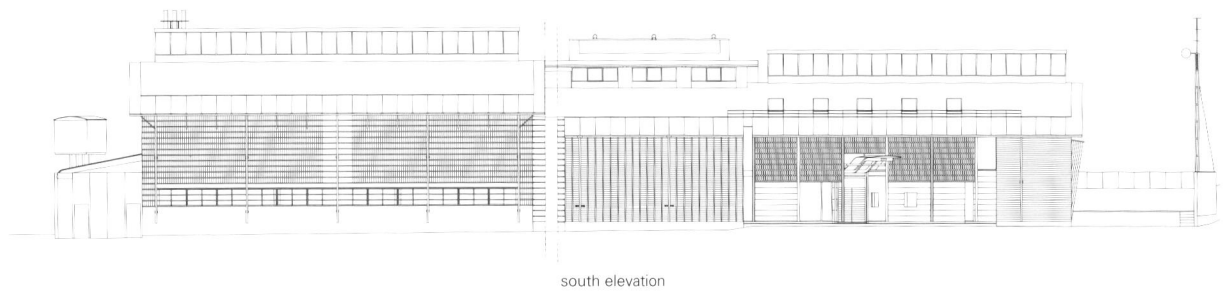
south elevation

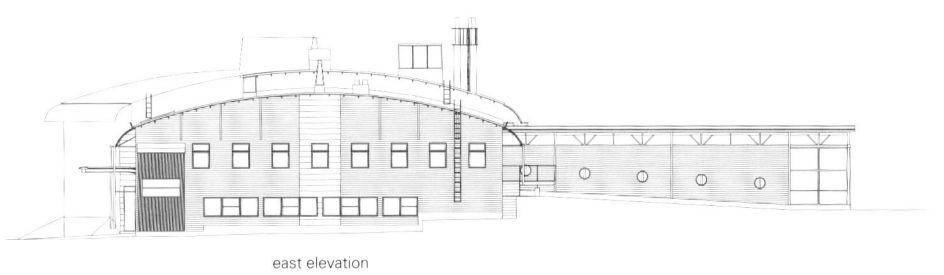
east elevation

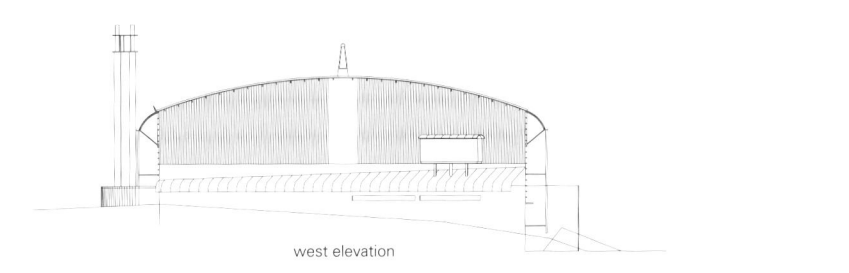
west elevation

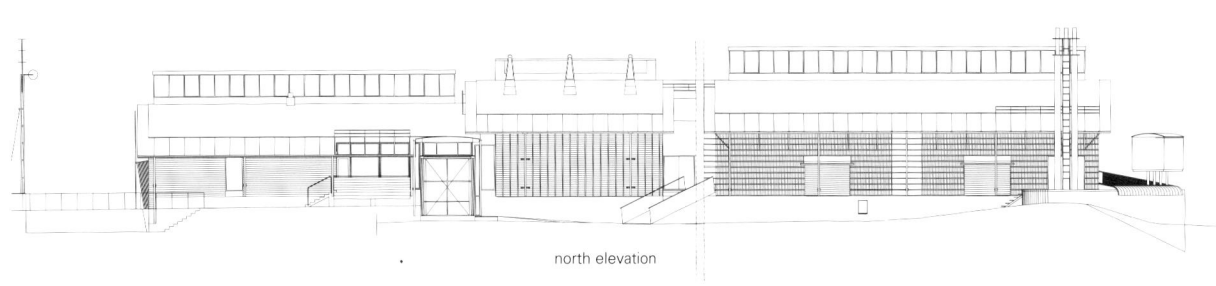
north elevation

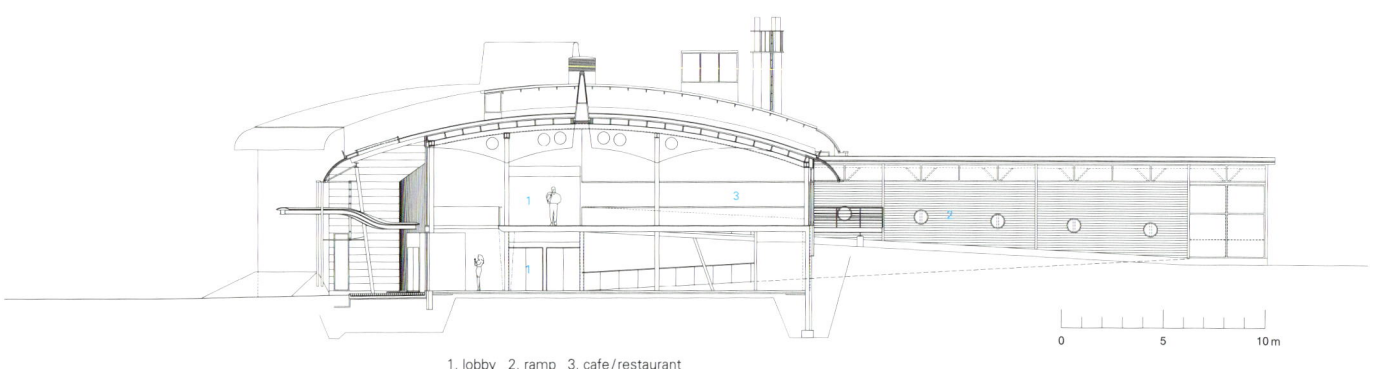

1. lobby 2. ramp 3. cafe/restaurant

건축

사미 박물관은 북극권에서 북쪽으로 300km 떨어진 이나리 마을의 끝자락으로 흐르는 강을 끼고 있다. 2층 규모의 박물관 건물은 1959년 설립된 사미 박물관 야외 전시관의 남서쪽 경계선에 위치한다. 과거 자갈 채석장이던 건물 대지는 척박한 대지의 모습을 바꿔 놓았다. 기존 야외 전시관과 같은 레벨에 있는 박물관 건물의 주 층은 출입구 로비 층에서부터 완만하게 경사를 이루는 램프를 통해 진입한다. 램프가 회전하는 지점에서 야외 전시관의 가장 인상적인 공간인 중정을 볼 수 있다. 지하층에는 입구 로비, 공연장, 임대전시실, 사무실, 방문객 안내센터, 창고, 작업실, 기계실 등이 있다. 최상층에는 소개전, 기행전, 북극권 생활문화 상설전 등을 위한 일반 전시실, 방문객들을 위한 카페테리아, 도서관, 사미 박물관 전용 사무실 등이 있다. 옥외 램프를 통해 야외 전시관으로 가는 길목에 매표소와 경비실이 있다.

수직, 수평의 요소가 분명한 박물관 건물의 외형은 풍경과 잘 어울린다. 지붕의 곡선형은 기복이 있는 주변지형을 반영하면서 한편 기나긴 겨울, 건물에 쌓이는 눈을 제거하기 위한 구조이기도 하다. 돌출한 곡선형 처마구조는 건물의 높이를 시각적으로 축소시켜 건물이 주변환경에서 도드라져 보이지 않도록 하는 역할을 한다. 건물 가운데 부분에는 통유리로 된 채광창이 있어 건물의 중심으로 햇빛을 모아들인다. 건물은 철근 콘크리트와 철골을 이용한 구조이며, 여러 가지 원목 패널을 외장재로 이용했다. 전시실은 짙은 남색이고 이외의 공간들은 모두 흰색으로 처리했으며, 적갈색의 원목은 밝은 공간과 대조를 이루기 위해 사용했다. 공공 공간에는 테라코타 안료를 혼합한 콘크리트가 바닥재로 이용되었고, 패널 거푸집을 이용한 바닥 아래쪽 기초부분까지 들여다보인다.

카페와 기계실의 조명을 제외한 일체의 조명기기는 박물관에 맞도록 주문제작한 것이다. 주요 층의 반간접 조명은 곡선형의 천장에 반사광을 드리운다.

전시공간

각 전시관의 구성은 전문가들로 구성된 특별위원회가 수년에 걸쳐 기획하여 준비했다. 건물과 전시실 설계의 일관성을 위해 동일한 건축가가 맡았다. 박물관 개관 1년 전에는 조명을 비롯한 기술적 솔루션을 시험해보기 위해 실제와 똑같은 모의 전시를 시행했었다.

전시실은 과학을 극장이나 미술 전시관의 경험적 측면과 접목시키는 데 주안점을 두고 설계했다. 소개전은 고대에서 현대에 이르는 일만 이천 년 동안의 북극지방의 문화, 생태학적 발전사를 보여준다. 북부 라플란드의 자연경관과 사미라프족의 문화를 보여주는 주전시관은 문화와 생태의 상호작용이라는 컨셉을 지니고 있다. 혹한에서의 삶의 형태와 방식을 계절의 변화에 따른 원형 구조로 보여주고 있다. 관람객들은 원형의 전시실을 돌아보면서 사미의 삶의 방식과 이곳의 자연현상이 1년을 주기로 어떻게 변화하는지를 살펴볼 수 있다. 원형의 바깥쪽에서 안쪽으로 들어가면 사미의 자연 및 문화유산, 현재 사미의 민족정신 사이의 상호작용과 연결고리를 읽어낼 수 있다. 사미 특유의 삶의 방식을 보여주는 자연환경의 초대형 사진 - 뒷부분에 조명이 비친다 - 들이 전시실의 바깥벽들을 장식하고 있어 라플란드의 자연, 기후, 빛의 변화를 세세히 보여준다. 이러한 배경 이미지에 사진과 드로잉, 해설, 동영상 등이 덧붙여져 있고, 월별로 기본 기후정보 및 어린이들을 위한 탐구기획 특집물도 있다. '자연과 문화' 전 사이의 인터페이스는 자연과 문화 공통의 주제를 지닌 전시구역이다. 원의 바깥쪽에서부터 관람하면 특정 자연현상을 알 수 있으나, 전시관 중앙바닥에서부터 관람하면 문화적 맥락이 읽어진다. 중앙바닥에는 뒤쪽에서 조명을 받는 사진들이 있는데, 사진 위에는 사미 특유의 삶을 묘사한 미술작품들이 설치되어 있다. 전시공간은 기본적으로 다소 어둡고 전시물에만 조명이 집중된다. 천장에 있는 채광창은 전시공간을 라플란드의 두 계절인 여름과 겨울을 대변하는 두 공간으로 분할한다.

전시공간을 둘러보면 흡사 자연을 거니는 것처럼 다채로운 경이로움을 맛볼 수 있다. 사진과 드로잉, 전시물과 해설의 효과를 증폭시키기 위해 계절 및 자연환경에 적절한 음향을 곁들였다. 관람객의 움직임에 의해 더 극대화되는 이러한 음향효과 역시 자연 속을 거닐면서 느낄 수 있는 감각적 경험을 떠올리게 한다.

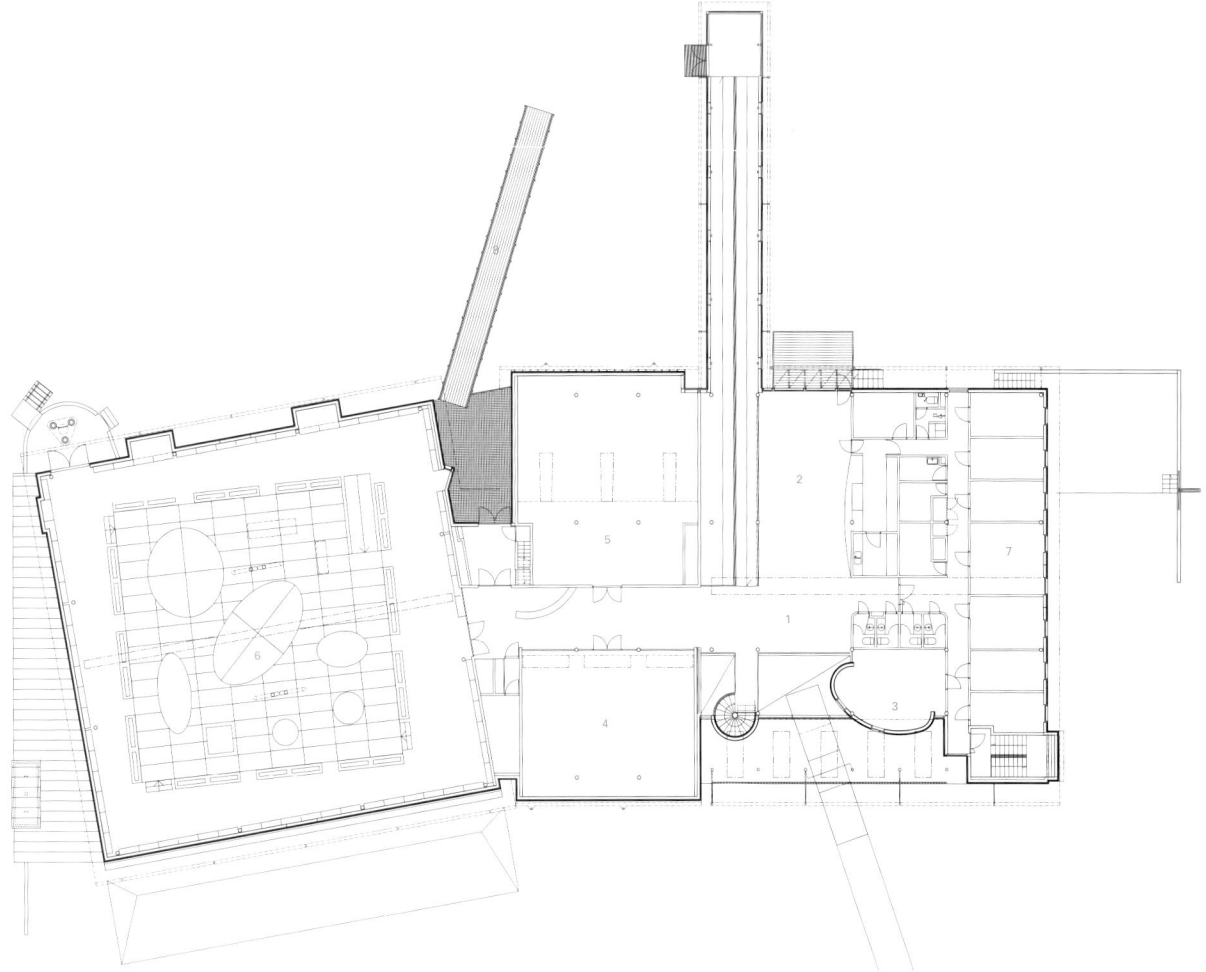

1. upper lobby 2. cafe/restaurant 3. library 4. introductory exhibitions
5. temporary exhibitions 6. permanent exhibitions 7. office 8. ramp to the outdoor museum

second floor

1. entrance lobby 2. guidance exhibition 3. ramp
4. auditorium 5. temporary exhibitions 6. storage and technical facilities

first floor

Architect: Juhani Pallasmaa Architects
Exhibition design: Juhani Pallasmaa
Architects - Juhani Pallasmaa, Sami Wirkkala
Structural engineering:
Rovasuunnittelu Oy - Jaakko Hamari
HVAC engineering: LVI - insinööritoimisto
Vidén Oy
Electrical engineering:
Sähköinsinööritoimisto Esko Laakso Oy
General Contractor: Levirakennus Oy
Client: Ministery of Culture
Location: Inarintie 1, Inari Finland
Total floor area: 2,590m²
Cubic volume: 11,700m³
Photograp: Rauno Träskelin

Museum of the Iron

철 박물관

Bianchini & Lusiardi associates
비안띠니 앤 루시아르디

This the interior design for a museum located in the Alps, in nothern Italy ; the museum deals with the production of the iron and its derivates, which so far has characterized this geographical area for the last five centuries. One peculiar aspect of this museum is that one of the most important objects in exhibition is the building itself, which is a huge melting and forging factory dating at the 16th century and exceptionally well preserved in its structure as well as in its ancient equipments.

For these reasons, our design has been conceived of as a bit different from usual.

Together with the scientific curators, we decided to find a way to take the inner spirit of the building back to life, to tell its story as well as the one of people who lived and worked in the factory during so many years, and to do this through subtle, evocative elements.

For these reasons, the interior setup of the museum is based on two levels of perception: the descriptive path, which is formed by panels, video projections and other elements telling the visitors the history of the building as well as the productions that were made in it during the centuries; and an evocative level, conceived of to suggest visitors the hidden spirit of the place in a more inconscious way, playing with visitor's senses and emotions. These two levels accompany people throghout the whole exhibition path, continously playing toghether and with the environment.

Some basic choices has been decided immediately, after the first sight at the future museum building.

For example, due to the complexity of the space and its appereance, we decided to use few simple materials yet with a strong appeal, like Cor-Ten steel and clear glass. Cor-Teen steel, wich has been used for almost all the metal elements, has many interesting qualities : it recalls the feeling of the old iron but at the same time is a contemporary, modern material ; furthermore, by gradually changing its color and texture with time, it suggests that the museum is a live organism, constantly trasforming and evolving.

Again the interior design has been conceived of as a mix of strongly physical elements and digital projections, to create a dialog between material and immaterial and between the new interior design and the old stone building; also the lighting has been designed to emotionally underline the building, its massive stone walls as well as its deep hollows.

The museum has been organized on three floors, with various spaces which include exhibition rooms, a conference room, a main video projections room, a ticket and book shop, laboratories and various service rooms, on a overall surface of about 1,500 m². It is also used for temporary exhibitions, conferences and seminaries mainly about industrial archeology.

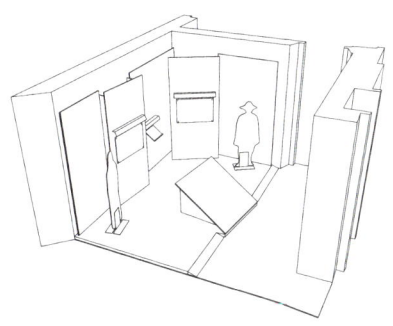

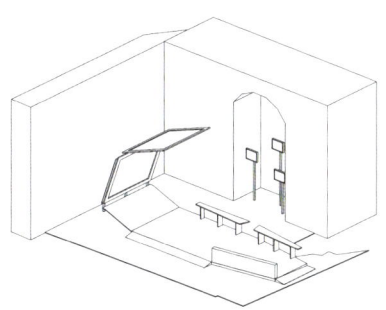

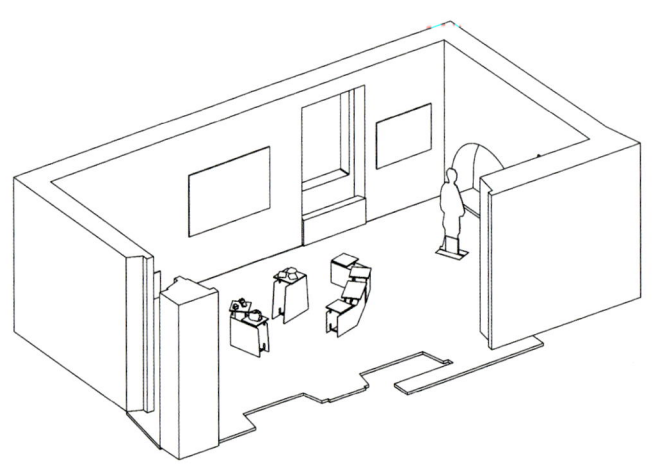

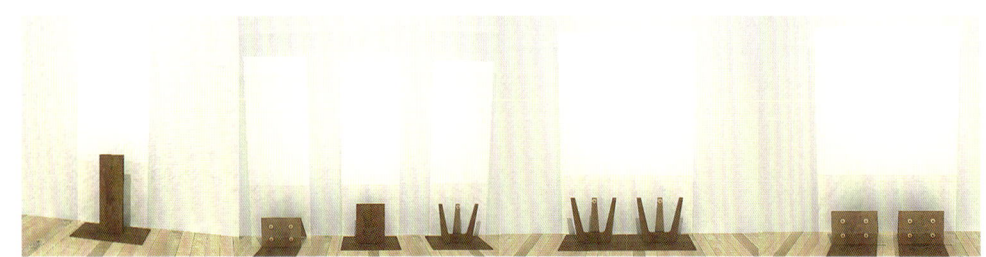
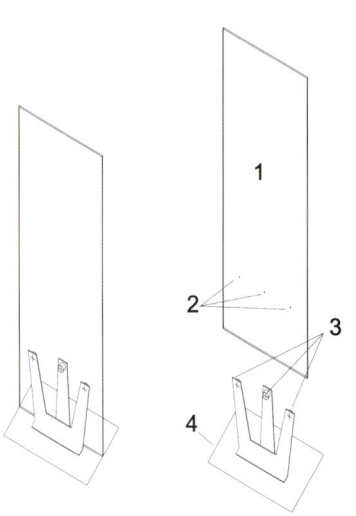

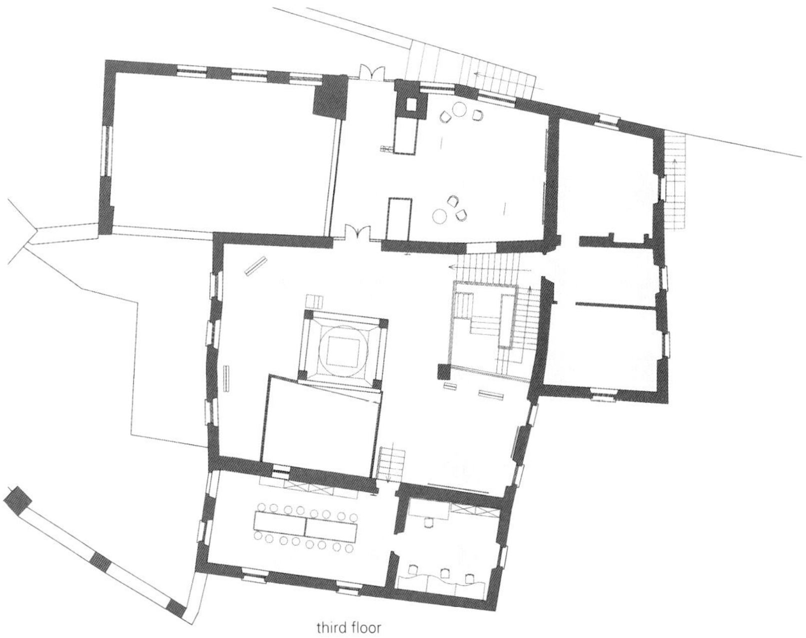

third floor

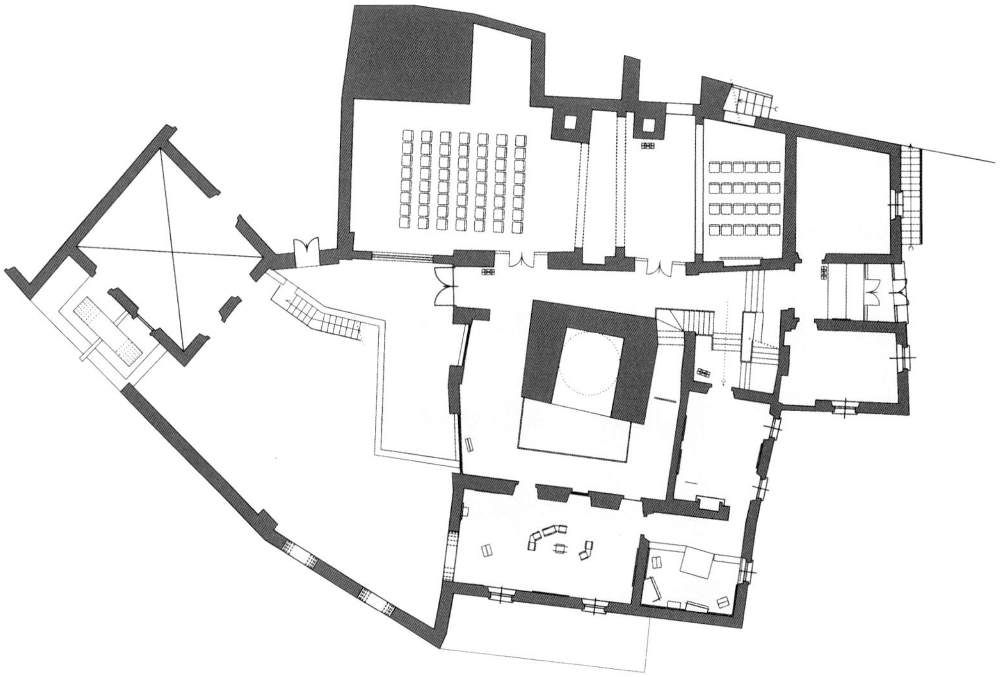

second floor

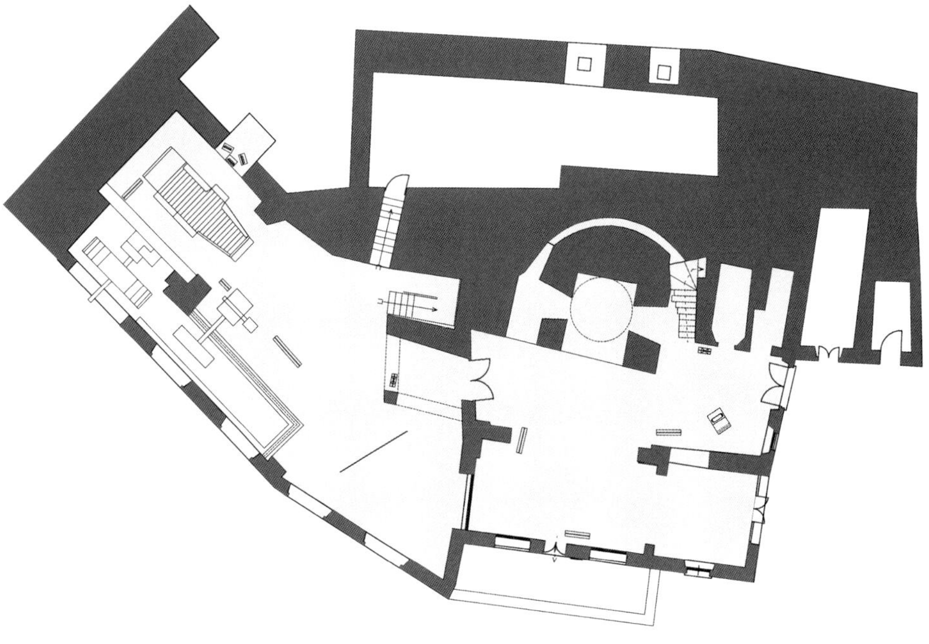

first floor

이태리 북부 지역의 알프스 산맥에 위치한 박물관을 위한 인테리어 디자인이다. 이 박물관은 지난 5세기 동안 이 지역의 특징이었던 철강 및 그 관련 제품들을 전시한다. 이 박물관의 특징 중 하나는 전시되는 가장 중요한 오브제 중 하나가 박물관 건물 그 자체라는 점이다. 박물관 건물은 16세기에 지어진 거대한 제철소이며 그 구조는 물론 옛 장비들이 매우 잘 보존되어 있다. 이 때문에, 우리의 디자인은 일반적인 디자인과는 조금 다르게 고안되었다.

우리는 합리적인 사고를 가진 큐레이터들과 함께 섬세한 역사적인 흔적들을 통해 건물의 내적 정기를 다시 불러일으키고 제철소의 역사를 설명하며 오랜 세월 동안 여기에서 일했던 사람들을 설명할 수 있는 방법을 찾기로 결정했다.

따라서 박물관의 인테리어 구성은 두 가지 차원에 근거한다. 하나는 방문객들에게 건물의 역사는 물론 그곳에서 수세기 동안 만들어진 제품들을 소개하는 판넬, 영상 장치 및 여타 요소들로 구성되는 설명 경로이고, 다른 하나는 방문객들에게 이곳의 숨은 정기를 보다 무의식적인 방법으로 보여줌으로써 그들의 감각과 정서를 자극하는 흔적을 찾는 것이다. 이들 두 차원은 서로, 환경과 상호작용하면서 사람들을 전체 전시 경로로 유도한다.

앞으로 박물관이 될 건물을 처음 본 후 몇 가지 기본적인 선택들이 바로 결정되었다. 박물관의 공간 자체와 그 모습이 복잡하기 때문에 우리는 코르텐 강철과 투명유리와 같은 눈에 잘 띄는 몇 가지 단순한 재료들을 사용하기로 결정했다. 모든 금속 요소들을 위해 사용되는 코르텐 강철은 흥미로운 특징들을 가지고 있다. 이 재료는 오래된 철과 같은 느낌을 주나 현대적인 재료이다. 또한 시간이 지남에 따라 그 색과 질감이 서서히 변하기 때문에 박물관이 끊임없이 변모하고 발전하는 살아있는 유기체처럼 보일 수 있다. 또한, 가시적인 것과 비가시적인 것, 그리고 새로운 내부 디자인 요소와 오래된 석조 건물의 대화를 촉진하기 위해 인테리어 디자인은 분명한 물리적 요소들과 디지털 영상의 혼합이다. 그리고 건물과 그 육중한 석벽은 물론 그 깊은 빈 속을 정서적으로 강조할 수 있는 조명을 설계하였다.

전시실, 회의실, 주영상실, 매표소 및 서점, 실험실과 여러 서비스 공간 등을 포함한 다양한 공간들이 3개 층에 구성되어 있으며 전체 면적은 약 1,500m²이다. 또한 박물관은 산업 고고학에 관한 임시 전시장, 회의장, 세미나실 등으로도 사용된다.

Designer : Bianchini e Lusiardi associati
Head of historical commetee : mr. Carlo Simoni
Main Contractor : Formenuove srl
Client : City of Tavernole sul Mella
Location : Tavernole sul Mella, Brescia, Italy
Size : 1,500 sqm on 3 levels
Description of the intervention : overall interior setup, rooms design, furniture design, lighting design, multimedia exhibit design, graphic design
Materials used : Cor-Ten steel, Glass, Larch wood, MDF, Polycarbonate
Photograph : Studio Pini

Dynamic Earth
에든버러 지구과학 전시관

Michael Hopkins
마이클 홉킨스

Edinburgh's much-admired urban character comes from a balance between nature and artifice, between the geology of its magnificent setting and the way generations of architects have refined its topography to create one of Europe's most elegant cities. The striking form of Dynamic Earth adds to the tradition. Its smooth, flowing curves and taut, minimal cables and masts offset the rugged majesty of Arthur's Seat and Salisbury Crags.

Just as the building's appearance suggests a relationship between nature and artifice, so its purpose helps visitors to understand that relationship. Located on the exact spot where James Hutton, the father of modern geology, lived and worked in the 18th century, it is an appropriate site for the Dynamic Earth exhibition. Interactive, virtual and wide screen film technologies can simulate any event in the Earth's evolution, from the Big Bang to the future. Visitors can experience earthquakes, volcanoes and the processes which formed natural settings like Edinburgh's.

The design helps visitors to relate to the grandeur of natural forces, which the exhibition illustrates, to the species of the site and its human history. It comprises three main parts : the fabric roof and its structure, a main two storey building which contains the exhibition, offices and workshops, and a forecourt. The first makes a generous entrance pavilion which looks towards the hills and the city ; although weathertight, its form and glass walls give the feel of an outdoor space. A hemispherical dome to a multi-media theatre bursts into the space, indicating something below. Two other structures rise into the pavilion, both containing staircases and lifts to the exhibition and

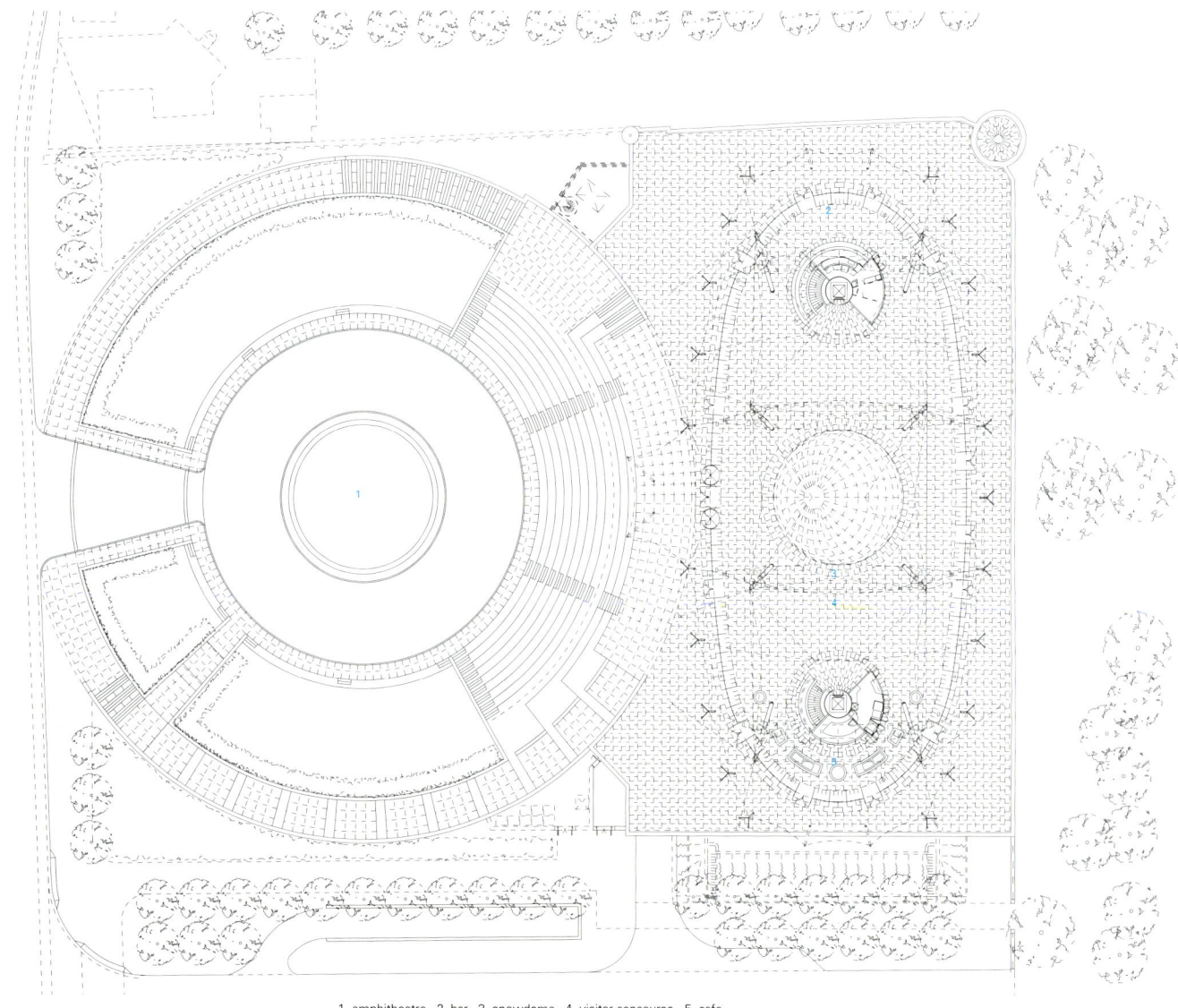

1. amphitheatre 2. bar 3. snowdome 4. visitor concourse 5. cafe

catering facilities for visitors who might want to linger.
The two storey exhibition space is a black box to give ideal viewing conditions. Offices and workshops are on either side of it. An old wall, which once formed the edge of a brewery, is restored and extended to make the external wall of the black box, suggesting in microcosm Edinburgh's fruitful relationship between geology and human intervention. The third component is a monumental forecourt, a new public space adding to Edinburgh's legacy of generous civic design. It's ampitheatre can be used for outdoor performances during the Edinburgh Festival; just as its neighbours, the venerable Holyrood Palace and the new Parliament building use architecture to suggest political evolution, so the Dynamic Earth Project uses architecture to connect natural history to human and civic life.

Client : Dynamic Earth Charitable Trust
Location : Edinburgh, Scotland
Site area : 15,000m²
Photograph : Keith Hunter

에든버러 시는 유럽 최고의 품격을 가진 도시 중 하나로 만들기 위해 장엄한 자연환경과 오랜 세월동안 건축가들에 의해 재정립된 지형 사이에서 자연과 인공의 균형을 찾았다. 에든버러 지구과학 전시관의 독특한 형태는 이곳 특유의 전통에 더해질 것이다. 매끈한 유선형 곡선, 그리고 간결하고 절제된 케이블과 마스트는 아더의 자태와 솔즈베리의 험한 바위산의 거친 위엄을 상쇄시킨다.

건물의 외관이 자연과 인공의 관계를 암시하는 것과 마찬가지로 건물의 목적 또한 방문객들이 그 관계를 이해하도록 정보를 제공한다. 현대 지질학의 아버지로 불리는 제임스 휴튼의 18세기 주택 겸 연구소가 있던 자리에 에든버러 지구과학 전시관이 들어섰으니 이보다 더 안성맞춤인 장소는 없을 것이다. 상호작용하는 와이드 스크린 필름 기술을 이용해 빅 뱅에서 미지의 미래에 이르기까지 지구 진화과정을 방문객들에게 시뮬레이션으로 보여준다. 방문객들은 지진이나 화산, 에든버러에 형성된 자연의 진화과정들을 가상으로 체험할 수 있다.

전시관은 관람객들이 위대한 자연의 힘을 이해하고 대지의 특성 및 인류의 역사를 파악하기 쉽도록 다자인했다. 전시관은 직물을 이용한 막구조 지붕과 전시관, 사무실, 작업실이 있는 두 개 층의 본관, 옥외공간의 세 부분으로 구성되어 있다. 지붕과 구조는 험한 바위산과 에든버러 시를 향하고 있는 널찍한 출입 홀을 구성하고 있다. 비바람을 막아줄 수 있는 견고한 구조인데도 건물의 유연한 형태와 유리벽은 옥외공간에 있는 듯한 느낌을 준다. 멀티미디어 극장이 위치한 반구형 돔은 아래쪽에 있는 뭔가를 암시하는 듯하다. 전시관에 우뚝 서 있는 두 개의 구조체에는 부대시설과 전시실로 접근할 수 있는 계단실과 승강기가 있다.

두 개 층에 걸쳐 있는 검은 박스형의 전시 공간은 전시물들을 둘러보기에 이상적이다. 전시관 양쪽에는 사무실과 작업실이 있다. 한때 양조장 벽으로 쓰였던 오래된 벽을 개조, 확장해 검은 박스의 외부 벽으로 활용하여 에든버러에서 볼 수 있는 지구와 인간의 공생적 관계를 은유적으로 보여준다. 건물 앞에 위치한 옥외공간은 공공 공간이 발달되어 있는 에든버러에서 또 하나의 공공장소로 자리매김할 것이다. 원형극장은 에든버러 페스티벌 때 야외공연장으로 활용될 수 있다. 인근에 있는 유명한 홀리로드 궁전과 신축한 국회의사당 건물이 이 곳의 정치적 변천사를 자연스레 암시하듯 에든버러 지구과학 전시관도 자연과 인간을 하나로 엮어줄 것이다.

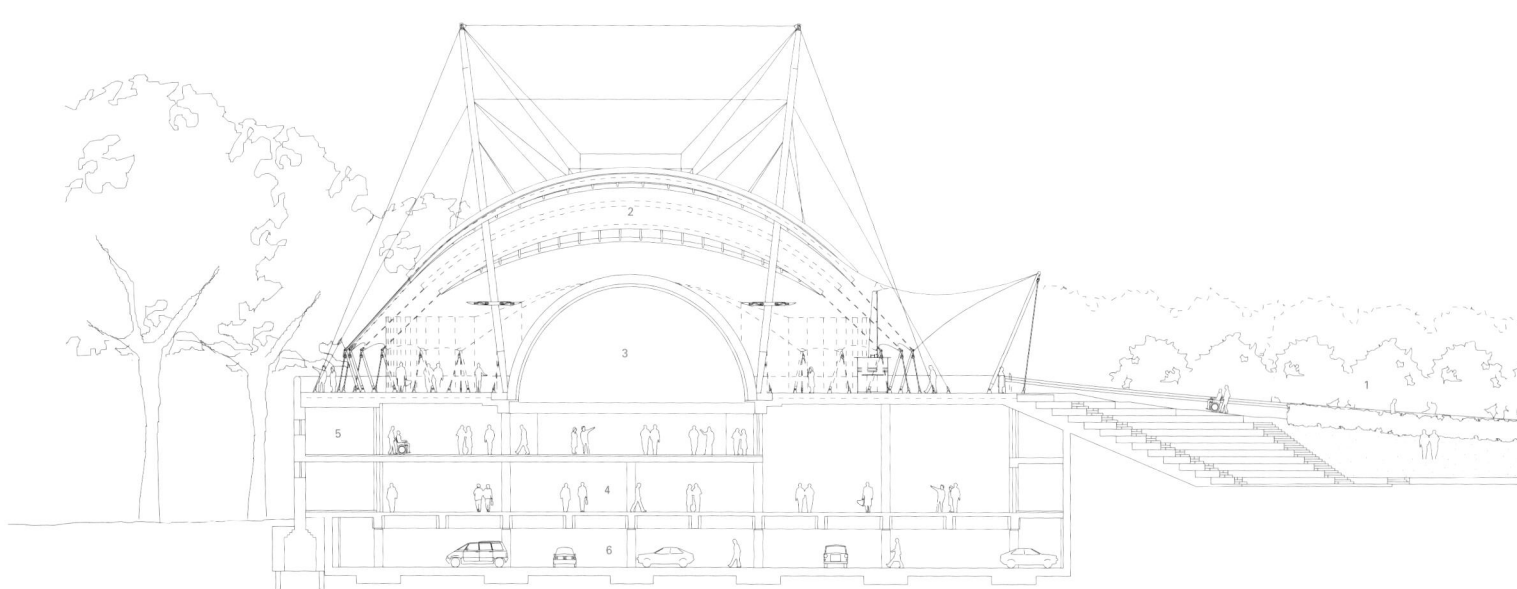

1. amphitheatre 2. visitor concourse 3. auditorium 4. exhibition 5. administration 6. car parking

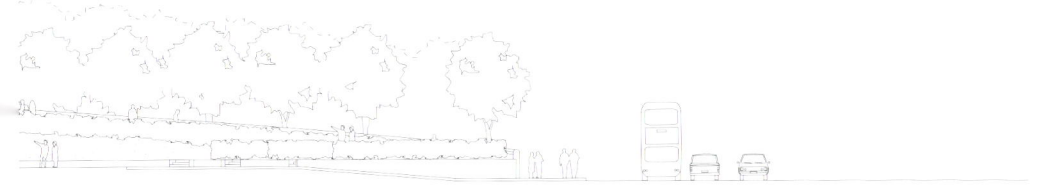

203

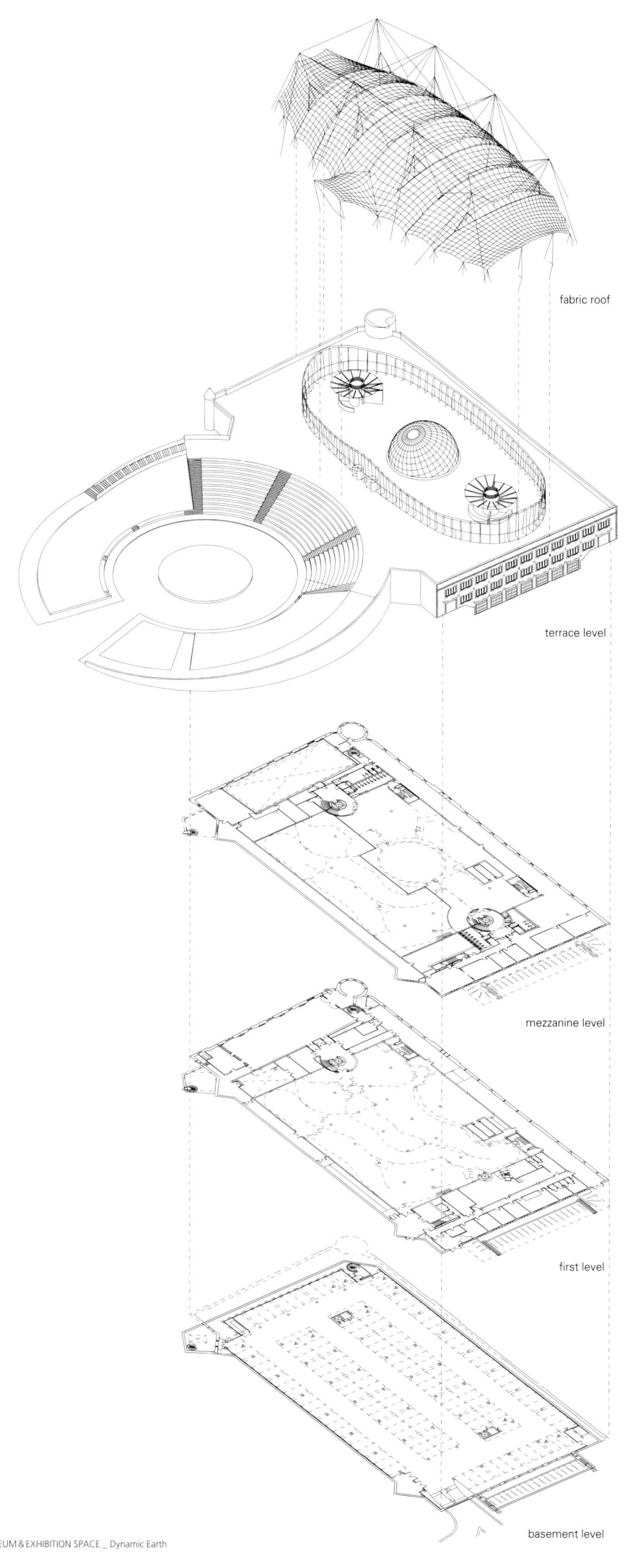

Hedge Building
담쟁이벽 전시관

Atelier Kempe Thill
아뜰리에 켐페 틸

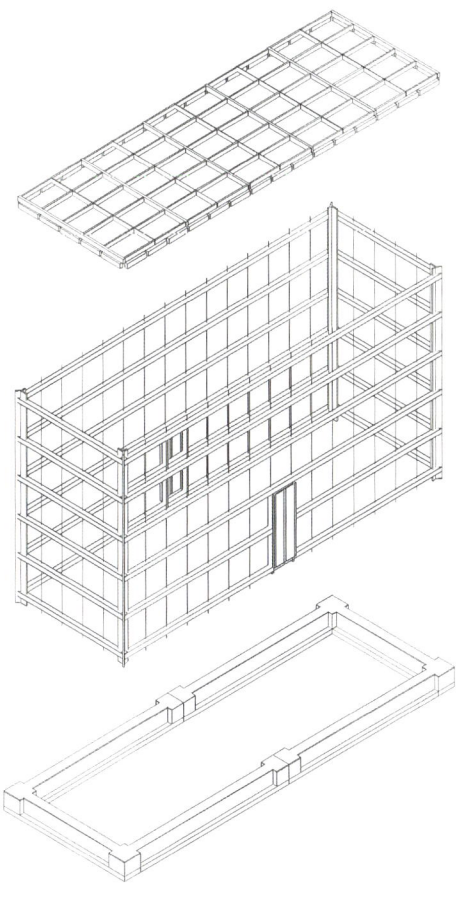

The project demonstrates the logic and rationality of Dutch agriculture and unearths surprisingly romantic qualities within its limited conditions. The architectural starting point is a fascinating new building element: the 'smart screen'. The smart screen is an ivy hedge grown in Dutch greenhouses. It is produced in sections measuring 1.2 by 1.8 metres and planted in gardens. Essentially an industrial product, the hedge can be deployed to build 'green walls'. Normally, it takes years for ivy to grow and cover a building. The smart screen makes a green building possible instantly.

The dimensions of the Hedge Pavilion are 20m x 6.5m x 10m. It is, in fact, a pergola. Its compact shape, topped by a roof and entered through four-metre-high doors, gives it the enclosed character of a house. This is balanced by the see-through character of the ivy. A steel framework creates five rows of channels filled with earth from which the smart screens grow. A computer-controlled system of pipes provides irrigation for the hedges. The structure has no conventional diagonal bracing but is stabilised by four star-shaped corner columns that can withstand all horizontal loads. Vertical loads are carried by a multitude of five-centimetre-thick columns. The hedges are visually continuous and look as if they support the structure. The hedges partly conceal the star-shaped columns, each of which weighs 4,000kg, and make them appear less substantial. Visually, the hedges seem to turn the corners, yet at the same time the corners are subtly marked.

The interior is enclosed on all sides by 10-metre-high green walls of ivy. Covering the top of the space is a screen of translucent plastic. The space is very neutral and modest and can accommodate different functions. Interesting light conditions dominate inside. Light entering through the ceiling gives the space the character of a classical museum, and more light filters through the enclosing hedges. The result is a play between inside and outside. The light

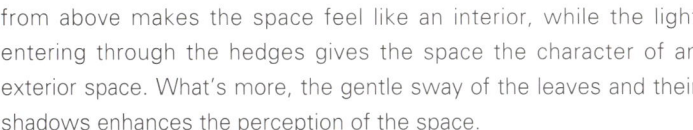

from above makes the space feel like an interior, while the light entering through the hedges gives the space the character of an exterior space. What's more, the gentle sway of the leaves and their shadows enhances the perception of the space.

이 건축물은 제한된 상황에서도 놀라울 만치 성공적인 네덜란드 농업의 합리성과 효율성을 단적으로 보여준다. 이 프로젝트의 주안점은 새롭고 매혹적인 건축요소인 '스마트 스크린'이다. 스마트 스크린은 네덜란드의 온실에서 볼 수 있는 담쟁이덩굴을 이용한 장벽이다. 벽의 기본 프레임은 1.2m x 1.8m 크기의 단면도를 조립해 정원에 설치한 것이다. 기본 프레임에 담쟁이덩굴을 조합하여 인공적으로 생산된 '초록의 벽'을 만들었다. 담쟁이덩굴이 자라나 건물을 뒤덮으려면 보통 수년이 걸린다. 이러한 단점을 보완한 스마트 스크린은 즉각적으로 그린 빌딩을 건축할 수 있게 한다.

전체 20m x 6.5m x 10m인 이 건물은 담쟁이덩굴이 뒤덮고 있는 파고라라 할 수 있다. 지붕이 있고 높이 4m의 출입문이 있는 이 건물의 간결한 형태는 사방이 막힌 주택의 특징을 지니고 있다. 반면 속이 투명하게 들여다보이는 담쟁이덩굴의 속성은 그러한 폐쇄성을 완화시킨다. 철골 프레임은 스마트 스크린을 성장시키기 위해 토양으로 채워진 다섯 줄의 수로를 만들었다. 이 구조에서 스마트 스크린이 뻗어 나온다. 파이프의 컴퓨터 제어 시스템은 이 장벽에 물을 공급한다. 전통적인 사선형 지주 대신 별 모양의 기둥 네 개가 코너에서 스마트 스크린을 떠받치고 있다. 수평으로 가해지는 일체의 하중은 코너 기둥 네 개가 지탱하고, 수직으로 가해지는 하중은 두께 5cm의 기둥 여러 개가 받는다. 담쟁이덩굴이 시각적으로 이어지면서 마치 철골 구조물을 떠받치고 있는 것처럼 보인다. 담쟁이 벽은 각각 4,000kg에 달하는 별 모양의 코너 기둥들이 담쟁이덩굴을 지지하고 있지만 그 중량감을 느낄 수는 없다. 시각적으로 담쟁이덩굴이 코너를 감싸고 돌기 때문에 약간의 흔적이 있을 뿐이다.

내부는 사면이 높이 10m의 담쟁이덩굴 벽으로 둘러싸여 있고 천장은 반투명 플라스틱 스크린으로 되어 있다. 중립적이고 차분한 이 플라스틱 천장은 여러 기능을 수용할 수 있다. 내부 조명 시스템도 흥미롭다. 천장을 통해 들어오는 햇빛 덕분에 고전적인 박물관의 느낌이 드는 한편 담쟁이덩굴을 통해 더 많은 햇빛이 사방에서 비춰든다. 건물 내부와 외부의 빛이 자연스럽게 교류하는 것이다. 천장에서 들어오는 빛은 내부 공간의 분위기를 연출하지만, 담쟁이덩굴을 통해 들어오는 빛은 외부 공간과 같은 느낌을 준다. 빛을 통해 담쟁이 잎과 그림자가 하늘거리면서 공간의 감각적인 느낌을 한층 더해준다.

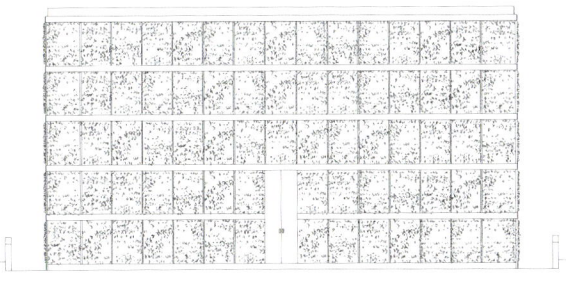

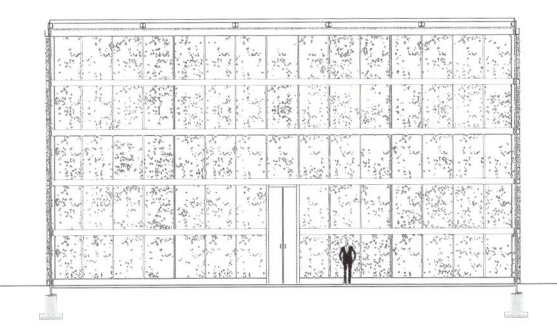

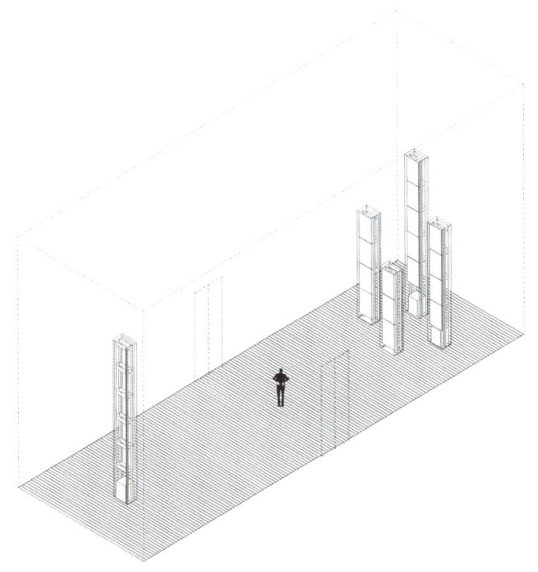

Architect: Atelier Kempe Thill architects and planners
Design team: André Kempe, Cornelia Sailer, Ruud Smeelen, Oliver Thill, Takashi Nakamura
Construction management: Matrix Architektur
Structural engineer: Rob Nijsse, ABT Velp
Landscape architect: Niek Roozen, Weesp
Video concept: Leo Schepman Communicatie Projecten, Den Haag
Smart screen: Mobilane, Nijmegen
Status: Dutch pavilion at the international garden exhibition IGA Rostock 2003 Germany, now used as cultural building by the city of Rostock
Size: 120m²
Photograph: Ulrich Schwarz

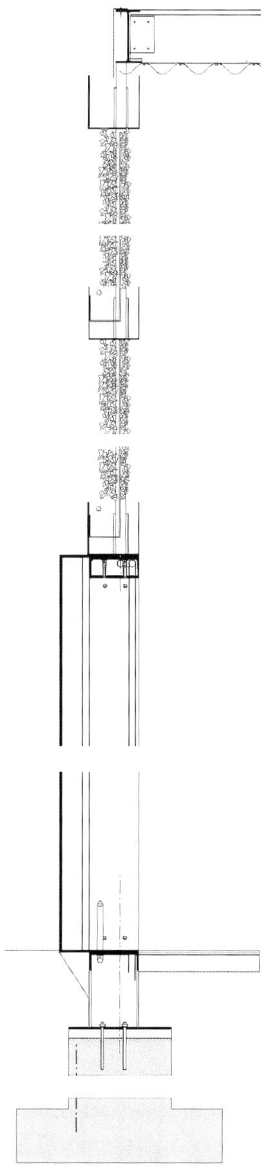

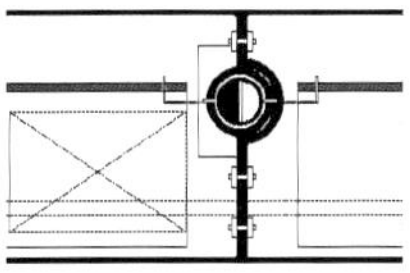

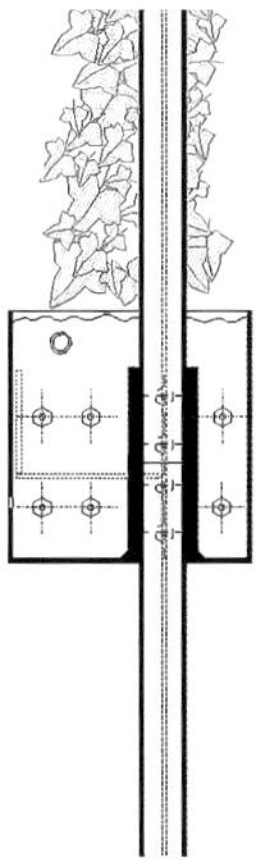

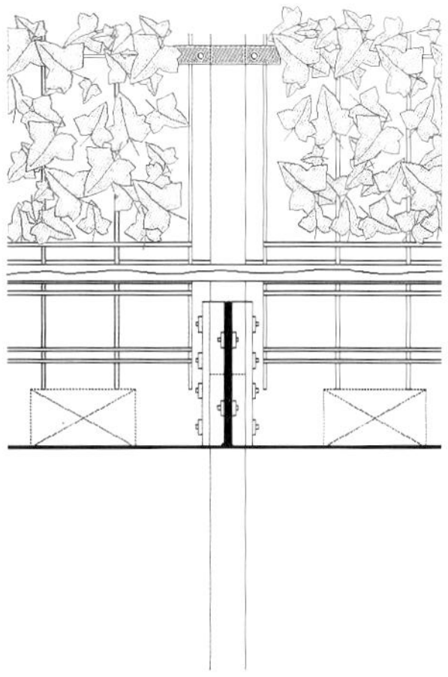

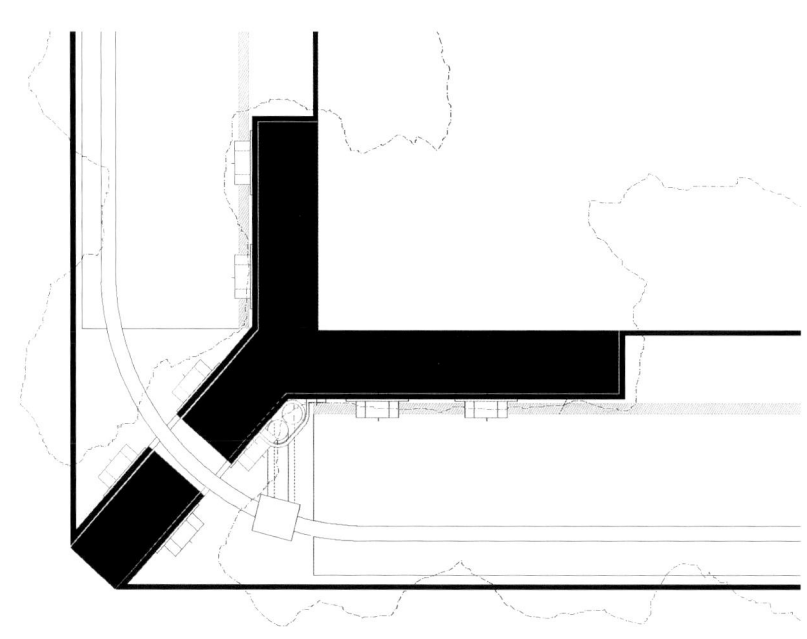

Inventaariokamari C74, Suomenlinna

인벤따리오까마리 C74

Arkkitehtitoimisto Laiho-Pulkkinen-Raunio
알끼떽띠또이마이스또 라이호 – 뿔끼넨 – 라우니오

The InventaariokamariInventory Chamber and its round log shed were built between 1778 and 1783 as an equipment store for the navy. It was probably designed by Fredrik af Champman.

In its original form, the Inventaariokamari was a three-storey building with a protruding middle section with a pediment. The simpler, hall-like round log shed also had three storeys. Both of the buildings were badly damaged in a fire started by the bombing in the Crimean War; only the exterior and partittion walls survived. They were later renovated as two-storey buildings, only the middle section remained three-storey. The main building was also converted into a provisions store.

An extensive renovation was carried out in 1868; the front wall and protruding front wall of the shed were dismantled, together with nearly all the partition walls made of stone. This made the building into a coherent whole. The facade on the Tykistölahti bay side also received its present appearance. The northern part of the building, the former round log shed, was destroyed in the bombing of 1944. The building was renovated by shortening the part next to the Inventaariokamari. Two wide door openings were also made.

The building has served as a store since Finland's independence. The Inventaariokamari has been renovated as an exhibition room and auditorium. The old part also houses storage space and an exhibition workshop. An annexe was built in place of the destroyed shed and now houses the Suomenlinna Info Centre.

Service areas, such as the ticket office, museum shop, and public toilets, are situated on the ground floor. The first floor houses staff facilities, offices and meeting rooms.

All technical spaces are located in the new part so that the old part could be left as untouched as possible. The elevations of the new part were made to be relatively solid following the style of the round log shed which was built in the Russian era, whereas the ceiling,

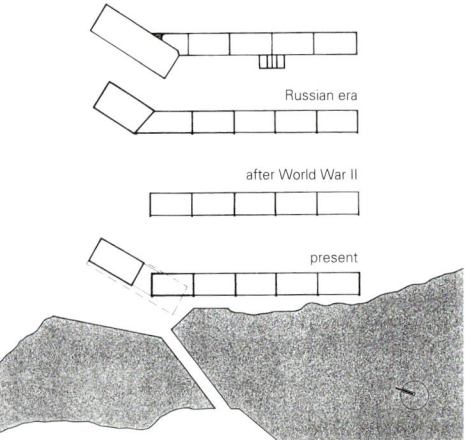

history of the Inventaariokamari

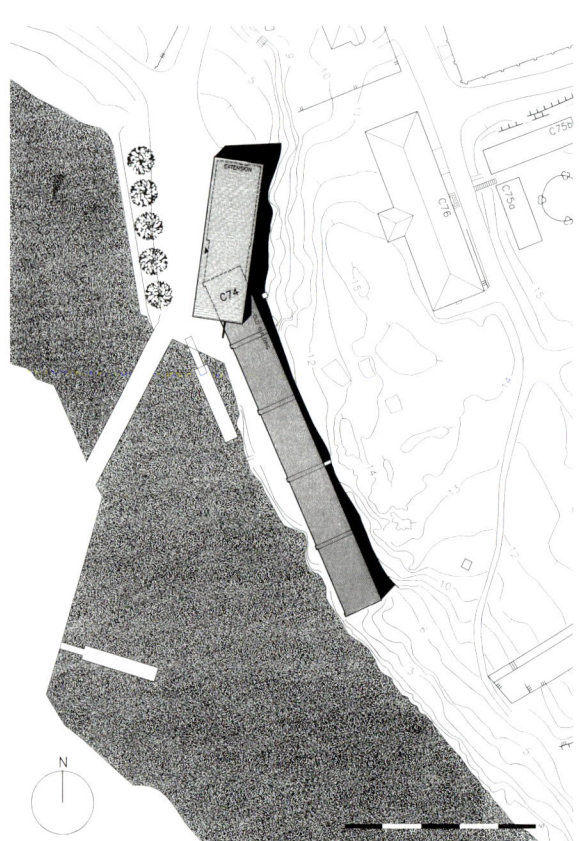

partly supported by columns, indicates the large size of the building in the Swedish era. The new and the old parts are connected by a glass articulation.

This articulation makes it easier to perceive the surviving part of the Inven -taariokamari and also makes the entrance hall of the Info Centre more spacious and interesting.

The new roof which also covers the old part ties the various construction phases and eras in with each other.

The loadbearing frame of the new building is a combined wood and steel construction. The timber beams are massive pine, recovered from the demolished building; the steel columns and beams are fireproofed L-shaped steel section. The exterior walls are of faced red brick. Inside the building, the brick, glass brick, steel, wood, and concrete surfaces have been kept as unaltered as possible. By constructing the offices as 'containers' on a steel structure, the new space has the feeling of the round log shed which used to serve as a store for ships' rigs.

In the Inventaariokamari exhibition of Suoenlinna's history, the panel frames are of warehousestyle perforated board to enable the flexible construction and alteration of the exhibition.

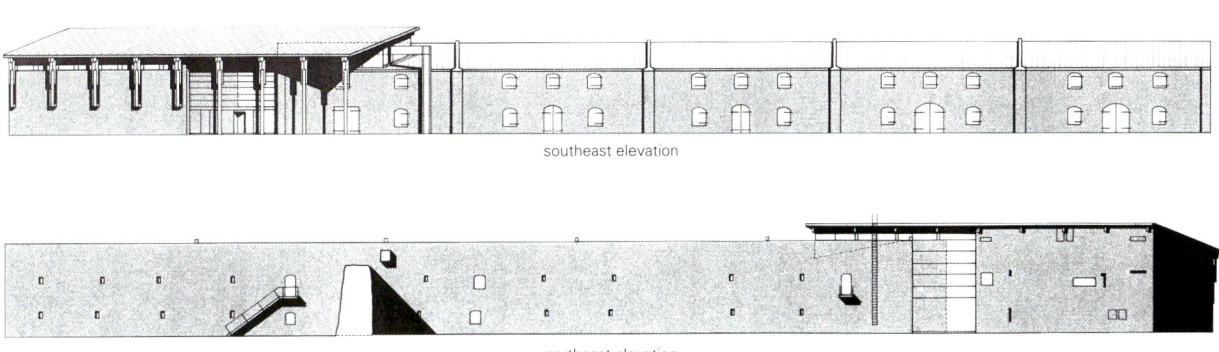

southeast elevation

northeast elevation

인벤따리오까마리와 원형의 통나무 건물은 1778〜1783년 사이에 해군 장비보관창고로 건립되었으며 프레드릭 챔프먼이 설계한 것으로 추정된다.

인벤따리오까마리는 원래 박공벽으로 중간부분이 돌출된 3층 건물이었다. 인벤따리오까마리보다 간결한 홀 같은 원형구조의 통나무창고 역시 3층이었다. 인벤따리오까마리와 통나무창고 모두 크림전쟁 때 폭격으로 인한 화재로 심하게 파손되어 외부와 파티션만 남았다. 그 후 돌출한 중간부분을 제외한 나머지는 2층으로 개조되었으며, 주 건물 역시 식료품점으로 개조됐다.

1868년 대대적인 개축공사에 착수, 건물의 앞쪽 벽과 돌로 만든 파티션의 대부분을 허물었다. 이로 인해 건물은 일관성을 갖춘 하나의 개체가 되었으며, 띠끼스또라띠만을 향하고 있는 파사드도 현재와 같은 외양을 갖추게 됐다. 원형 통나무 창고였던 건물 북쪽은 1944년 폭격으로 파괴되었으나, 인벤따리오까마리 옆부분을 짧게 만들어 개조했다. 개조하면서 두 개의 커다란 출입문을 신축했다.

이 건물은 핀란드 독립 이래 상점으로 이용되었다. 이후 건물은 전시실 겸 강당으로 개조되었으며, 남아 있는 옛 부분에는 창고와 전시용 워크숍으로 구성되었다. 파괴된 통나무창고 자리에 신축된 별관에는 수오멘리나 안내센터가 들어서 있다.

매표소, 박물관 기념품점, 공중화장실 등의 공용구역은 1층에 자리잡고 있으며, 2층에는 직원 전용시설, 사무실, 회의실 등이 있다. 모든 기계실은 신축 건물에 배치하여 가능한 한 남아있는 옛 건물의 원형을 훼손하지 않도록 했다. 신축 건물의 입면을 러시아 통치시대에 건립된 원형 통나무창고 스타일의 영향으로 견고한 반면, 기둥이 부분적으로 떠받치고 있는 천장은 스웨덴 통치시대에 유행한 대형 건물을 연상케 한다. 새 건물과 옛 건물을 이어주는 유리 구조물을 통해 인벤따리오까마리의 남아 있는 부분이 또렷하게 보일 뿐 아니라 안내센터 출입구 홀도 보다 널찍하고 흥미롭게 보인다. 옛 건물까지 이어지는 신축 지붕은 이 건축의 다양한 구조와 시기를 이어주는 역할을 한다. 신축 건물의 하중을 받는 골격에는 원목과 강철을 결합한 구조를 이용했다. 철거한 건물에서 회수한 소나무를 신축 건물의 들보에 재활용했으며, 강철을 이용한 기둥 및 들보는 내연성의 L자형으로 했다. 외벽마감재로 붉은 벽돌을 이용했으며, 가능한 한 건물 내부의 벽돌, 유리벽돌, 강철, 원목, 콘크리트 표면 등의 원형을 훼손하지 않도록 했다. 사무실 설계를 강철 구조물 속의 '컨테이너' 개념으로 계획하였는데, 이로써 신축 건물이 해군 군함의 삭구창고로 이용되던 원형 통나무 건물의 느낌을 유지하고자 했다. 수오멘리나의 역사를 반영하고 있는 인벤따리오까마리 전시관에 창고 스타일의 구멍뚫린 판자를 패널 프레임으로 사용함으로써 융통성과 가변성을 지닌 전시관을 창출하고자 했다.

Antiquarian control : The National Board of Antiquities, Suomenlinna
Architectural design : Arkkitehtitoimisto Laiho-Pulkkinen-Raunio Oy - Mikko Pulkkinen, Tiitta Itkonen, Philip Kronqvist
Structural design : Insinööritoimisto Innostructura - Eero Kotkas
Mechanical and electrical design : Projectus Team Oy
Juha Aberg, Erkki Hakanen
Building contractor : Seicon-Rakennus Oy
Client/Governing Body of Suomenlinna, Ehrensvärd-Society,
The National Board of Antiquities
Bldg. area : Total/2,330m^2
addition/510m^2
renovated part/1,820m^2
Bldg. volume : Total/10,240m^3
addition/3,240m^3
renovated part/7,000m^3
Photograph : Studio Voitto Niemelä

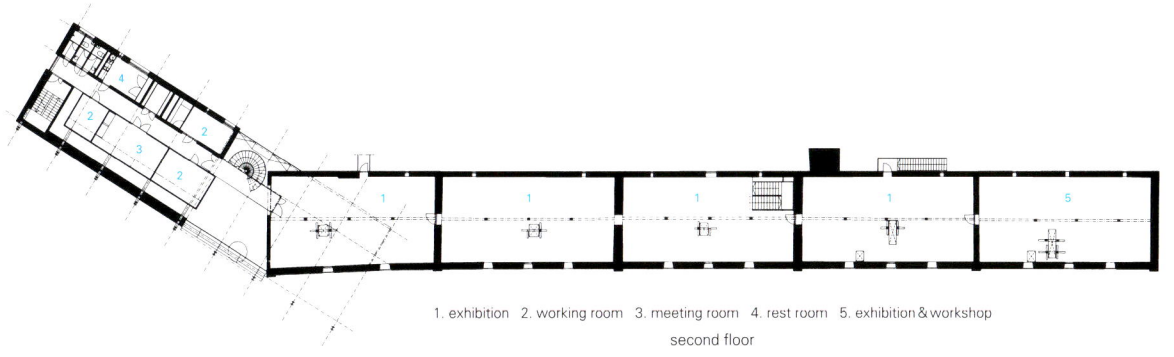

1. exhibition 2. working room 3. meeting room 4. rest room 5. exhibition & workshop
second floor

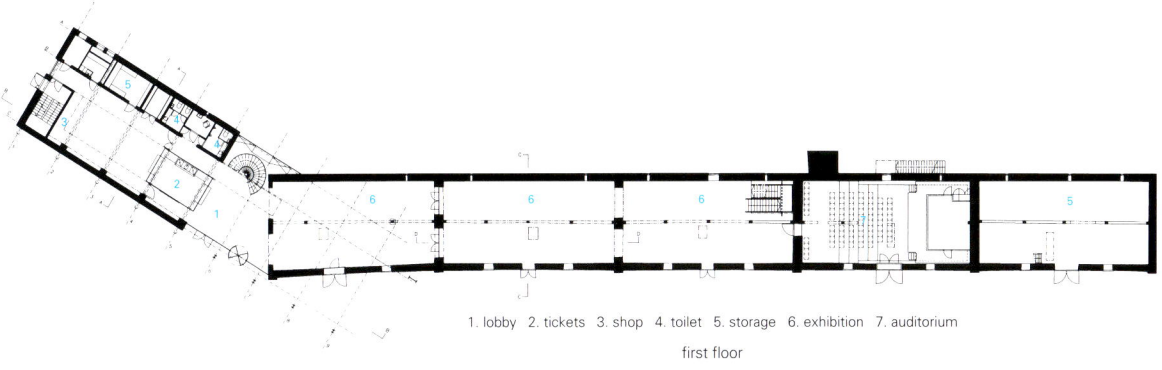

1. lobby 2. tickets 3. shop 4. toilet 5. storage 6. exhibition 7. auditorium
first floor

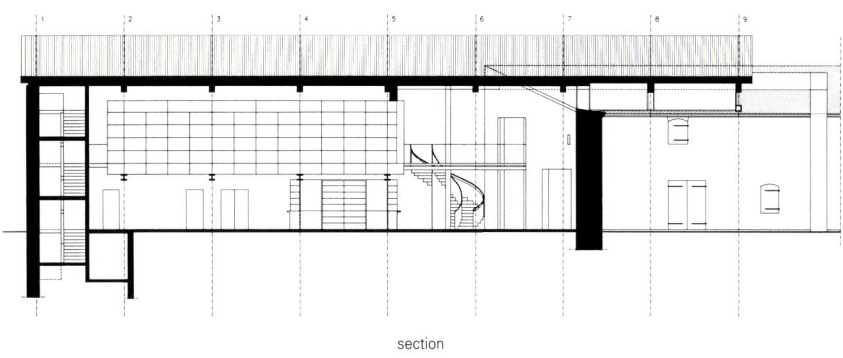

section

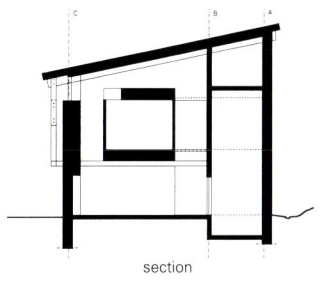
section

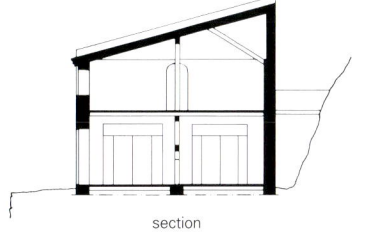
section

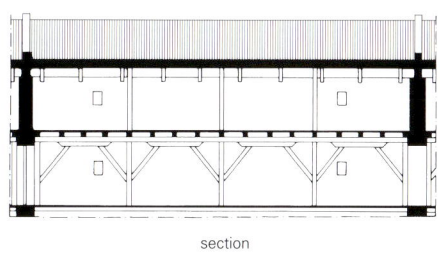
section

220 MUSEUM & EXHIBITION SPACE _ Inventaariokamari C74, Suomenlinna

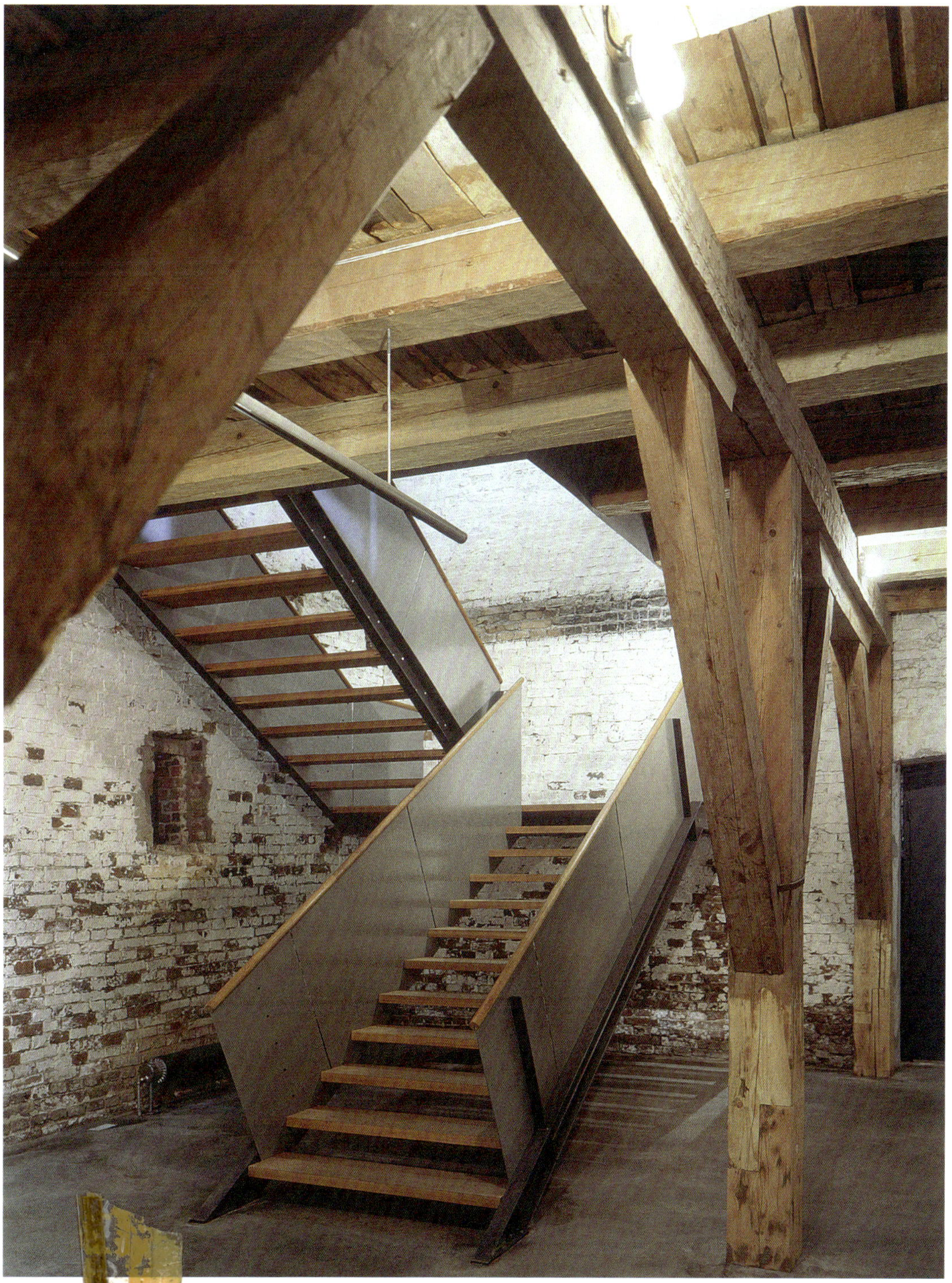

SUOMENLINNA 250
SVEABORG 250

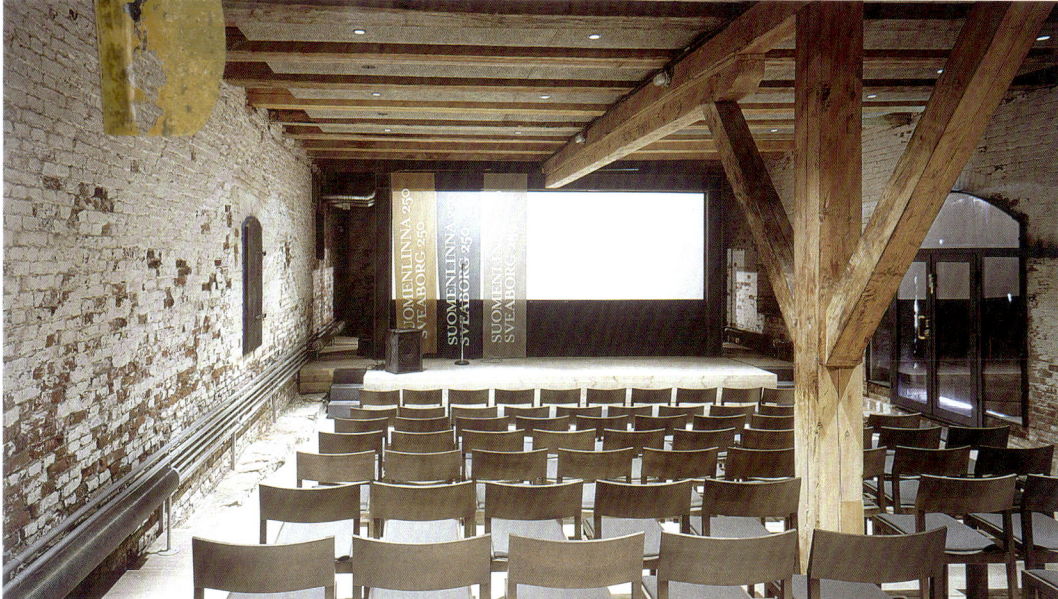

223

Rehabilitation of the Charles V Palace in the Alhambra

알함브라 찰스 5세 궁전

Juan Pablo Rodriguez Frade
후안 빠블로 로드리게스 프라데

Since 1527, the Charles V Palace has cut off on a bias one of the ends of the Courtyard of the Myrtles in Granada's Alhambra, instilling its stark, cubic geometry in contrast to the sensual spirit of the labyrinth-like Nazari architecture.

One culture imposes itself upon another, although it does so with a certain degree respect and admiration for the work of the Arabs. Fatality and even paradox cast their shadows on this building, considered to be one of Spain's and Europe's most prized renaissance jewels.

Done by Pedro Machuca, it has ended up serving as a showcase for Islamic art. The centuries passed and the Palace never saw completion, as the style of itinerant courts making this type of structure necessary died with the death of Emperor Charles V. As of the reign of Phillip II, the capital of Spain was established in Madrid and interest in completing Machuca's work gradually waned.

In photograps from roughly 1920, one could see the building as merely a set of stone walls without even a roof to cover them. History's legacy of an un-finished palace served as a starting point for architect Juan Pablo Rodríguez Frade to formalize his proposal for a project to turn the building into the Alhambra Museum.

The rehabilitation project stems from a thorough analysis of documents and photogrammetic surveys of each and every one of the stages in the building construction.

This not only provides a study of the changes made over history, it also allows for the elimination of all additions made during this century detracting from the palace's early essence by hiding its original structures. Closed up windows, false joisting, inner walls and other additions were removed, leaving the stone masonry in view as living proof of the monument's history.

From there on in, the elements needed for the functioning of the museum were respectfully added, almost tiptoeing in, barely disturbing the monument, slipping in between the marble flooring and the walls and casting their almost carpet-like covering concealing the lighting and heating and the cooling system, and, most importantly, heihtening the moveable nature of all of the project's elements. This not only guarantees a clearly differentiated reading of the new and old, but also provides reversibility, leaving any other type of future use for the building unhampered.

No mimicry or formal linkage to the palace that could eventually confuse the viewer is attempted whatsoever in this design. Contemporary language is therefore used with great concision here, establishing a soft, discrete dialogue between the old and new yet with striking, timeless materials such as iron and wood.

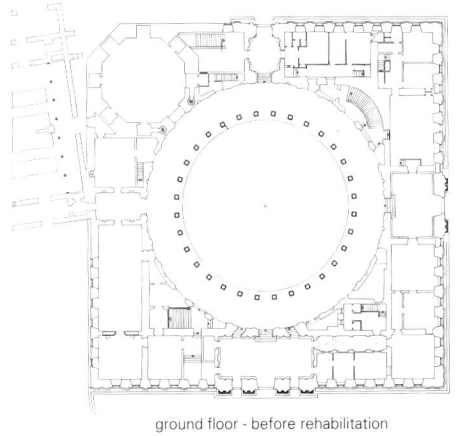
ground floor - before rehabilitation

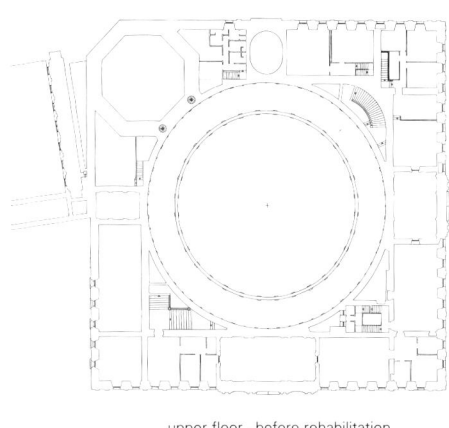
upper floor - before rehabilitation

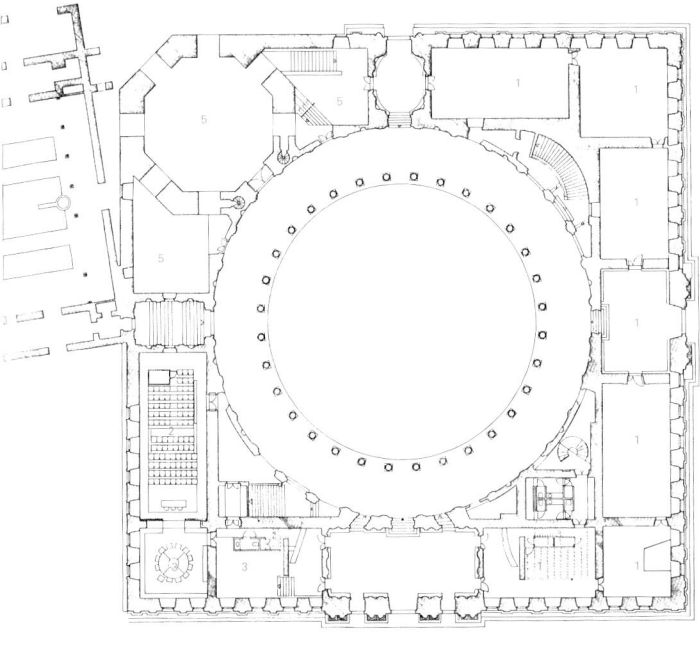
1. museum 2. auditorium 3. reception hall & meeting room
4. patio 5. temporary exhibition
ground floor

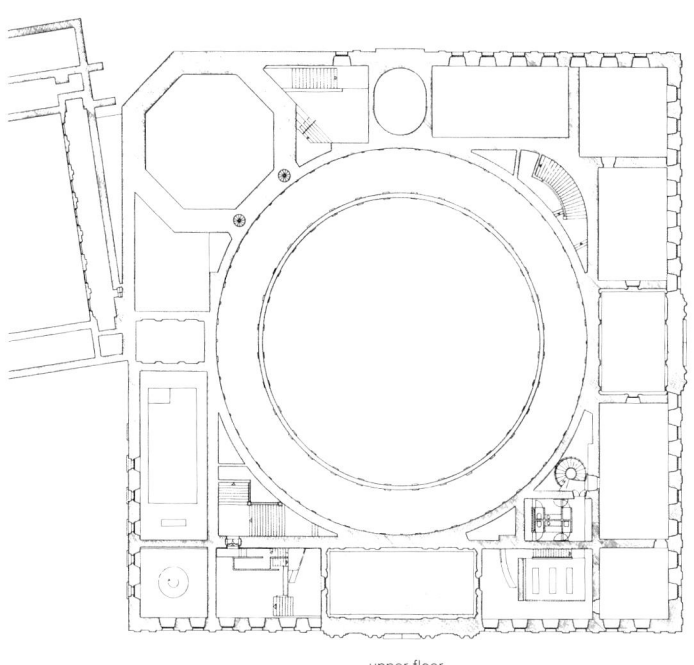
upper floor

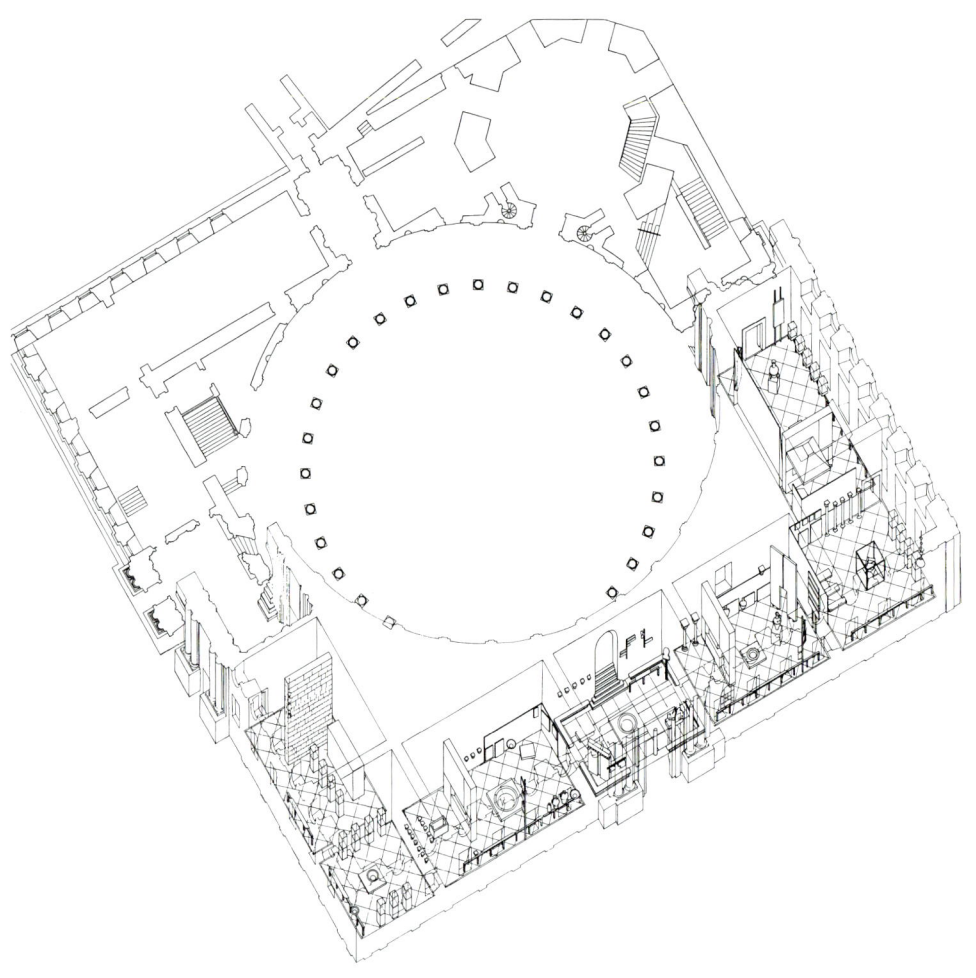

226 MUSEUM & EXHIBITION SPACE _ Rehabilitation of the Charles V Palace in the Alhambra

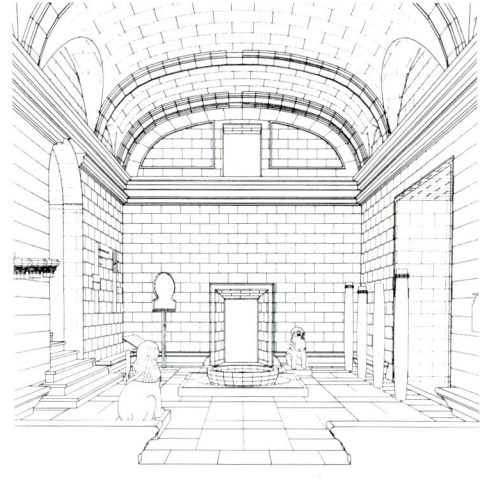

230 MUSEUM & EXHIBITION SPACE _ Rehabilitation of the Charles V Palace in the Alhambra

1527년 은매화정원의 비스듬한 한쪽 끝이 소실되면서 그라나다 알함브라의 찰스 5세 궁전은 휑한 입방체형으로 남아 미로 같은 나자리 건축의 감각적인 정신과 뚜렷한 대조를 이루어 왔다.

비록 다른 문화 속에 자리잡고 있지만 이 궁전은 아랍인들의 예술작품에 대한 일말의 존경심과 경탄을 간직하고 있다. 그러나 스페인과 유럽을 통틀어 가장 탁월한 르네상스시대의 보석으로 간주되는 이 궁전에는 불운한 운명과 나아가 역설의 그림자가 드리워 있다.

페드로 마추카가 건축한 이 궁전은 결국 회교 미술품의 전시공간이 되고 말았다. 그리고 수세기가 지나면서도 궁전은 완공되지 못하고 이러한 유형의 건축이 필요했던 궁전 스타일도 찰스 5세 황제의 서거와 더불어 사라졌다. 필립 2세가 스페인의 수도를 마드리드로 정하면서 마추카의 거작 완공에 대한 관심도 점차 희비해졌다.

1920년대까지만 해도 그 당시 촬영한 사진들을 보면 이 궁전은 지붕조차 없이 돌벽만 덩그러니 서있는 황량한 건물에 불과했다. 이에 건축가인 후안 빠블로 로드리게스 프라데가 미완의 유적으로 남아있던 이 궁전을 알함브라박물관으로 개조하겠다는 제안서를 제출했다.

개축 프로젝트를 위해 그는 일체의 궁전 건축 과정 및 관련서류를 철저하게 조사, 분석했다. 그런 조사작업을 통해 오랜 세월에 걸쳐 이뤄진 개축작업을 꼼꼼히 조사함은 물론 궁전의 원래 구조와 본질을 저해하는 금세기의 증축부분을 제거할 수 있었다. 폐쇄 창문, 모조 들보, 내벽 등의 추가물들을 제거함으로써 유구한 역사를 증명하는 석조건물을 원래 그대로 복원할 수 있었다.

이런 복원작업을 거친 후 원형을 훼손하지 않는 범위 안에서 대리석 바닥과 벽 사이에 박물관 기능에 필요한 요소들을 조심스럽게 추가했다. 카펫과 같은 부드러운 자재 안에 조명, 냉난방 시스템을 설치하고, 무엇보다 박물관에 부합하도록 모든 요소가 유동성을 가질 수 있도록 했다. 이를 통해 옛 것과 새 것의 차별화된 해석이 가능해질 뿐 아니라 향후 새로운 용도로 건물을 재활용할 수 있도록 배려했다.

건물을 설계하면서 궁전을 모방한다거나 궁전과의 어떠한 연계도 모색하지 않음으로써 박물관 관람객을 혼란시킬 요소들을 철저히 배제했다. 현대의 건축언어를 이용할 때는 간결함에 초점을 맞춰 옛 것과 새 것 사이에 원만하고도 신중한 대화를 유도했다. 아울러 마감재로서의 특징이 뚜렷하고 시대성에 구애받지 않는 철과 목재를 주로 사용했다.

Collaborators : Angel Cruz Plaza,
Javier García-Vaquero Alvaro
Work collaborator : Luciano Rodrigo Marhuenda
Quantity surveyors : Juan de Dios López Cantos,
Angel Aparicio Olea
Historical consultants : Pedro Galera Andreu,
Jesús Bermúdez López
Structure : Alfonso Gómez Gaite
Installations : Antonio Medina, GEASYT, S.A.
Coordination & supervision : Virginia Bueno Bueno,
Jorge Calancha de Pasos
Construction company : OCP Construcciones
Exhibition mounting building : BEC, S.A.
Lighting : ERCO
Financial management : Patronato de la Alhambra y
Generalif, Consejería de Culture,
Junta de Andaiucía, SOGEFINSA
Location : La Alhambra, Granada, Spain
Photograph : Lluis Casals

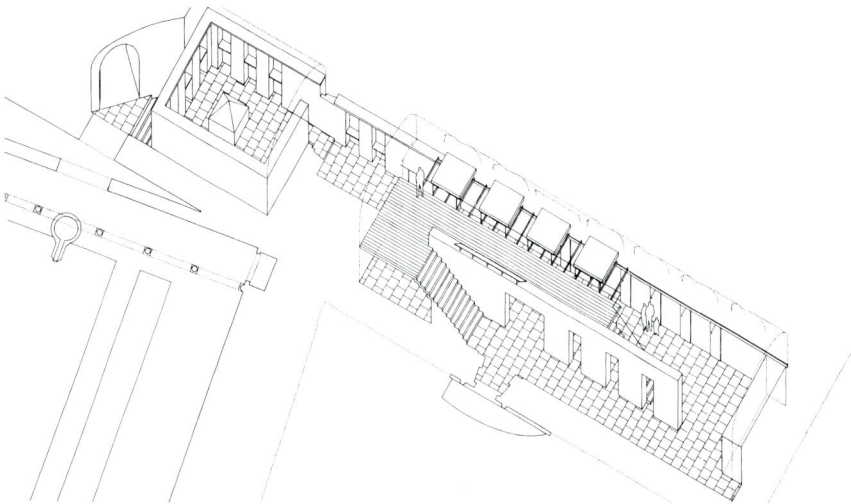

232 MUSEUM & EXHIBITION SPACE _ Rehabilitation of the Charles V Palace in the Alhambra

Invisible Box
투명상자

GAD
GAD

Esma sultan is a multi-purpose event space in Ortakoy, Istanbul, in the center of the city in a busy entertainment district on the banks of the Bosphorus.

Planting a glass and steel box inside the ruins of a palace to create a covered venue that constantly reminds us of the multiple histories innate in the new design.

The brick palace was built approximately 200 years ago for Esma Sultan, an Ottoman Sultan's wife as a summer palace. Destroyed by fire over a century ago, the exterior brick walls are all that remain of the building. In 1999 The Marmara Hotel decided to adaptively reuse the beautiful land-marked ruin, keeping the walls as a framework and support for a modern interior space inserted within to create an event and exhibition venue. Commi-ssioned to renovate and redesign the space, gad designed a thin but strong stainless steel and glass box that is suspended within the brick structure.

Glass buildings are often inappropriate in countries that have hot climates, however, the brick surrounds that remain of the palace made it possible for GAD to introduce this rarely seen architecture in Turkey, to Istanbul. The brick walls inadvertently create a shelter for the transparent glass box from the sun, rain and wind.

Multi-leveled, the new building incorporates a bar and restaurant on the ground floor and a conference room or event space on the second floor entered by a wooden and steel curved staircase. The glass box is tethered to the brick walls with suspension rods, which ensures the two separate structures remain equidistant from each other and can therefore withstand extreme weather conditions and earthquakes.

From the outside, the building gives the illusion that the palace remains in its original state. From inside, guests are reminded of the building's former incarnation with views of the Bosphorus made possible though the original arched brick window frames. The building encourages comparison between modern construction methods with those of 200 years ago.

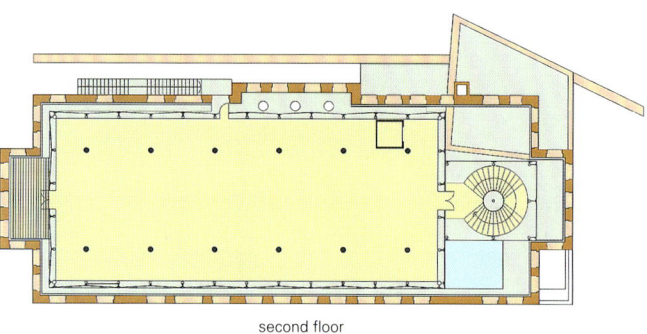

second floor

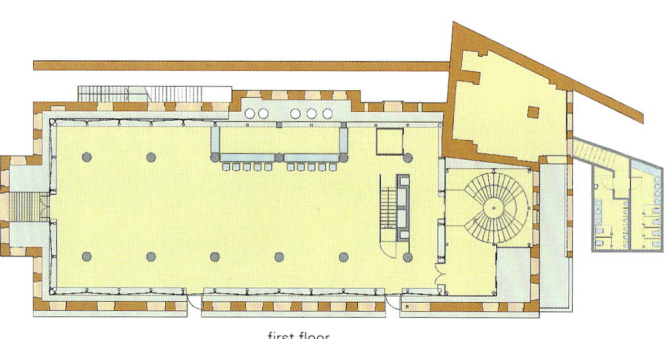

first floor

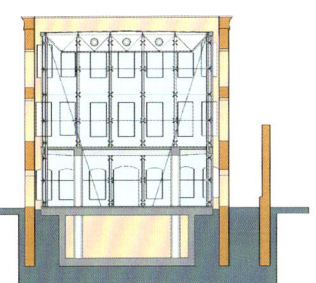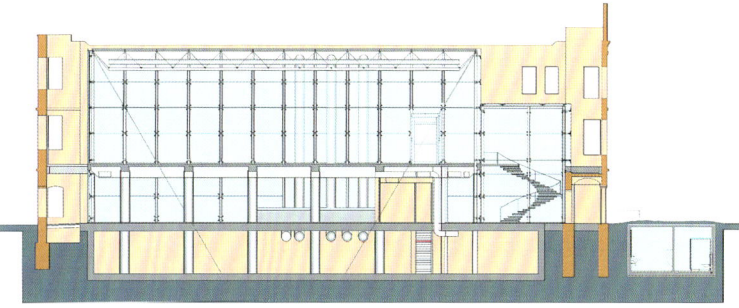

Architect: Gokhan Avcioglu (principal), Salih Kucuktuna
Assistant designer: Kerem Turker
Communication manager: Ozlem Ercil
3D images: Salih Kucuktuna
Location: Ortakoy, Istanbul-Turkey
Site area: 5,000m²
Bldg. area: 657m²
Total combined floor area: 1,492m²
Photograph: Courtesy of the architect

이스탄불 도심인 오르타쿄이에 자리잡고 있는 에스마 술탄은 다목적 이벤트 공간이다. 투명한 유리 상자 형태를 띠고 있는 신축건물과 옛 궁터가 자리잡고 있는 이 지역은 보스포러스 해협 연안의 번화한 엔터테인먼트 중심지다. 폐허가 된 옛 궁터 안에 들어선 이색적인 유리와 강철의 '투명상자'는 그곳의 파란만장한 역사를 대변하는 듯하다.

벽돌 궁전은 약 200년 전 오스만 제국의 어느 술탄의 아내였던 에스마 술탄의 여름 별장이었다. 100여 년 전 화재로 타버려 지금은 궁전의 외벽만 간신히 명맥을 유지하고 있다. 1999년 마르마라 호텔 측이 남아있는 궁전 외벽을 기본 골격으로 삼아 그 내부에 현대적인 공간을 신설, 이벤트 및 전시관으로 활용한다는 계획 아래 여전히 신비로운 아름다움을 간직하고 있는 유적지에 대한 대대적인 건축에 착수했다. 개수공사를 하고 공간을 새롭게 디자인하기 위해 얇지만 강도 높은 스테인리스 스틸과 유리를 이용한 박스 형태를 설계, 벽돌 궁터에 매달려 있는 형태를 취하도록 했다.

유리 건물들은 고온기후에는 적합하지 않으나 벽돌 구조가 유리를 에워싸고 있기 때문에 터키, 특히 이스탄불에선 좀처럼 찾아보기 힘든 유리 구조를 대담하게 선택했다. 벽돌 외벽들이 투명한 유리 상자를 강렬한 태양과 비, 바람으로부터 막아주는 역할을 톡톡히 할 수 있다.

신축건물의 지상층에는 바와 레스토랑이, 나무와 강철로 만든 곡선형 계단으로 이어지는 2층에는 회의실과 이벤트 공간이 있다. 유리 상자는 서스펜션 장치를 이용해 궁터의 벽돌 벽에 연결되어 있는데 이는 두 개의 분리된 구조가 서로 동일한 거리를 유지함으로써 터키의 폭염과 폭우, 지진을 이겨내도록 한 것이다.

외부에서 보면 유리 상자로 인해 옛 궁전이 마치 원래의 모습을 되찾은 듯한 착각을 불러일으킨다. 궁터의 아치형 벽돌 창문틀을 통해 유유히 흘러가는 보스포러스 해협을 내부에서 내다보면 궁전의 원래 모습이 머릿속에 떠오르는 것만 같다. 유리 상자를 보면서 우리는 현대적 건축양식과 200년 전 건축양식을 새삼 비교해보게 된다.

Gagosian Gallery
가고지안 갤러리

Gluckman Mayner Architects
그러크만 매이너 아키텍츠

Located in the Chelsea district of Manhattan, this 25,000sq.ft space has the scale of a small museum and the spatial versatility to accommodate a diverse range of exhibition requirements, including traditional and experimental installations. The program consists of a 2,400sq.ft Main Gallery, two smaller, skylit galleries, a Special Purpose Gallery, a Small Prints and Video Showroom, a Stock Showroom with 23-foot-high viewing walls, as well as professional office space. The raised roof provides the 6,000sq.ft, column-free, Long Term Installation Gallery with the capacity to house large-scale sculptures. A 10-foot-tall plastic clerestory bathes the space in natural light during the day, and at night, the lighting acts as a prominent 'sign' visible from the West Side Highway.

맨하탄의 첼시 지구에 위치한 가고지안 갤러리는 면적 25,000ft²로 는 아담한 규모에 비해 전통 설치미술과 실험적 설치미술 등 다양한 전시요건을 충족시킨다. 2,400ft²의 주 전시실과 소규모 전시실 두 곳, 특별기획 전시실, 시청각실, 10ft 높이의 관람용 벽이 있는 소장품 전시실, 직원용 사무실 등을 수용한다. 지붕이 높고 기둥이 없어서 초대형 조각품들을 전시할 수 있는 6,000ft²의 장기 설치전시실이 들어설 수 있었다. 낮에는 3m 높이에 있는 플라스틱 천창을 통해 자연채광을 받아들이며, 밤에는 멋진 갤러리 조명이 서부 고속도로에서도 눈길을 끄는 '간판'의 역할을 한다.

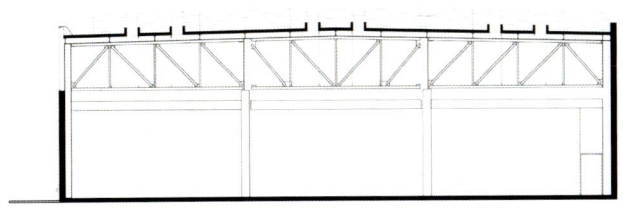
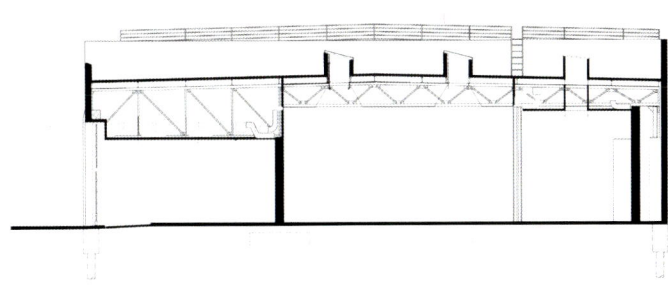

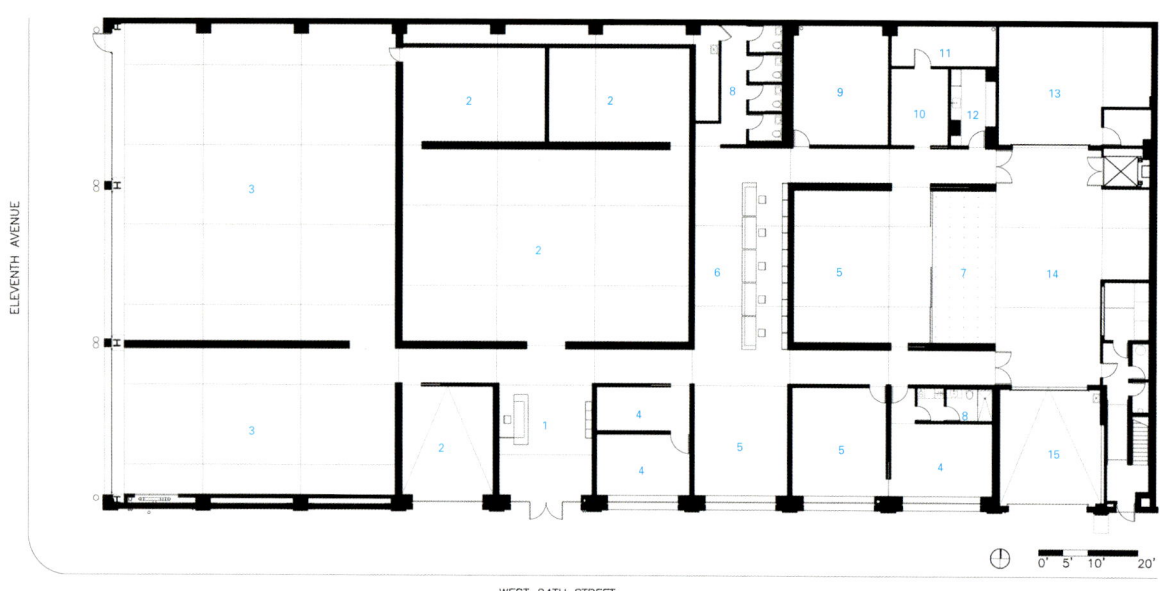

1. reception 2. gallery 3. long term installation 4. office 5. showroom 6. administration 7. racking 8. bathroom
9. print/video room 10. copy room 11. archive room 12. pantry 13. storage 14. workroom 15. art handling

floor

Principal architect: Richard Gluckman
Project manager: Michael Hamilton
Design team: Elena Cannon, Wilmay Choy, Ching-Loong Tai, Celia Cheng, David Pysh
Consultant: Leslie E. Robertson Associates, R.L.L.P. (structural engineer), Cosentini Associates LLP (mechanical engineer), Langan Engineering and Environmental Services, P.C. (geotech engineer)
Contractor: Eurostruct, Inc. (general contractor), Maspeth Welding, Inc. (structural steel), Urban Substructures (shoring), Air Force Mechanical Corp. (HVAC), Big Shot Electric Corp. (electrical), Citron Bros. Plumbing & Heating (plumbing/sprinkler), Atlantech Systems Inc. (skylight)
Client: Larry Gagosian
Location: 555 West 24th Street, NY, USA
Site characteristic: Existing brick garage (c.1940) in Chelsea Gallery District
Program: Long-term exhibition gallery[6,000sq.ft], Special purpose gallery[480sq.ft], Private showroom[760sq.ft with 23'-0"high walls], Small prints and video showroom
Site area: 25,000 sq.ft
Material: Terrazzo-ground concrete[floor], "Danpalon" polycarbonate by CPI International[skylights and clerestory], Anodized aluminum and sandblasted glass by Arm-r-lite[storefront], Brick with precast concrete base and sills[exterior facade], Resawn Yellow Pine[furniture and shelving]
Photograph: Harry Zernike

Wind Nest the Finnish Pavilion
바람의 안식처

SARC Architects
SARC 아키텍츠

Finland participates in the EXPO 2000 with its own pavillion. The pavillion is built to serve as a permanent building and its design is based on the winning proposal in an architectural prize contest. The pavillion is situated at a central location in the Eastern Pavillion Area between the east-west Europa Boulevard and Neue Latzener Strasse Süd.

The heart of the pavillion is a Finnish birch grove with erratic boulders. The Pavillion is entered through the birch grove which is intensively perceptible during the entire tour in the Pavillion. The trees are 12 to 15 meters of height and the ground is covered with a rich, scenting grass stand. The birch grove is separated from its noisy surroundings with partly silk-screened glass walls. From the outside the two four-storey pavillions look like monolithic pieces of wood lined with dark brown heat-treated birch.

The exhibition rooms are situated on the first and second floors and a restaurant is located by the exit on the first floor. An outdoor area by the southern wall, partly covered with a light-structured sail roof, serves as an outdoor restaurant. The third floor will host a flexible office area. More offices, the necessary engineering and utility services rooms as well as a sauna with adjacent lounge area for 6 to 10 people will be situated on the fourth floor.

The design is ecological by being flexible. After the exhibition the yard between the two pavillions can be covered with glass, and the pavillions are easily turned into offices. The placing of the pavillion on the site will enable an enlargement of both pavillions by 5m thus allowing the remaining permitted building volume to be utilized.

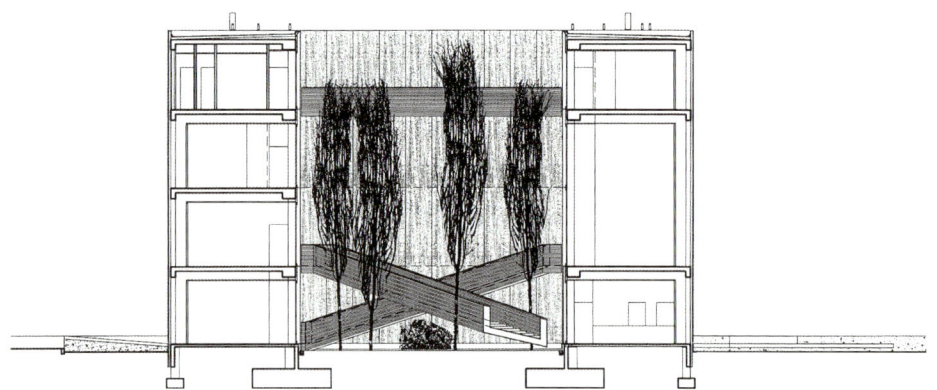
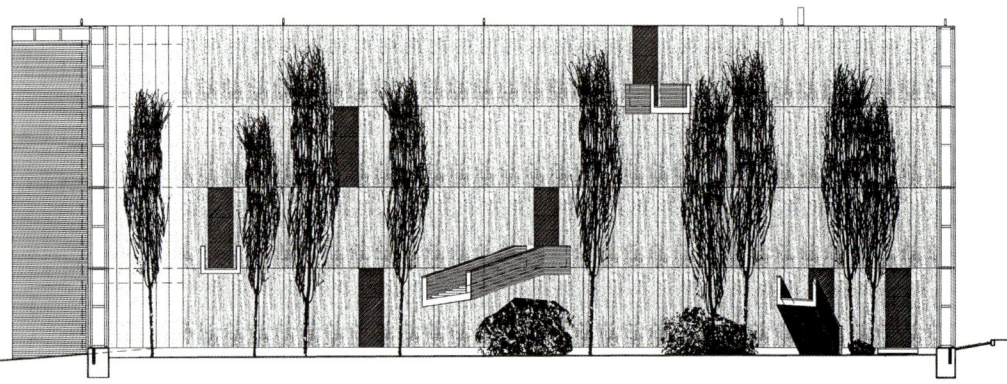

244 MUSEUM & EXHIBITION SPACE _ Wind Nest the Finnish Pavillion

핀란드는 2000년 세계 박람회의 국가관에 참가하게 되었다. 핀란드 파빌리온은 반영구적인 건물로 이용할 목적으로, 건축설계경기 입상작을 토대로 디자인했다. 이 파빌리온은 동서쪽에 위치한 유로파 대로와 신 라츠너 도로 사이에 있는 파빌리온 구역의 중앙에 자리잡고 있다.

불규칙한 형태의 자갈을 깔아놓은 핀란드 파빌리온의 심장부는 핀란드산 자작나무 숲으로 꾸며져 있다. 방문객들은 강렬하게 시선을 잡아당기는 높이 12~15m 정도 되는 자작나무 숲을 가로질러 파빌리온으로 들어간다. 부분적으로 실크 스크린으로 처리한 유리벽을 통해 자작나무 숲은 주변 소음으로부터 차단된다. 외부에서 보면 네 개 층의 전시관 두 곳이 암갈색 자작나무들로 되어있는 균일한 목재 구조물처럼 보인다.

1, 2층에 전시실들을 두고, 1층 출입구 근처에 레스토랑을 위치했다. 돛을 지붕으로 한 가벼운 구조가 부분적으로 덮고 있는 남쪽 벽으로 구분되는 외부구역은 야외 레스토랑의 역할을 한다. 3층에는 기능에 맞게 가변성을 가지는 사무공간이 있고, 4층에는 사무실, 각종 설비 및 기계실, 6~10인용 라운지가 딸려 있는 사우나 등이 들어설 예정이다.

파빌리온은 용도와 기능에 따라 유연성을 가지도록 생태학적으로 디자인했다. 전시가 끝난 후 두 파빌리온 사이의 야외 공간은 유리를 덮어서 쉽게 사무실로 개조해 사용할 수 있다. 파빌리온은 행사 후 5m까지 확장할 수 있도록 배치했는데, 이는 건축 최대 허용한계를 염두한 계획이다.

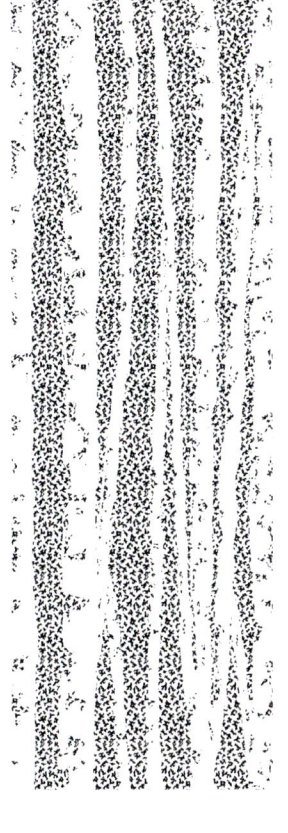

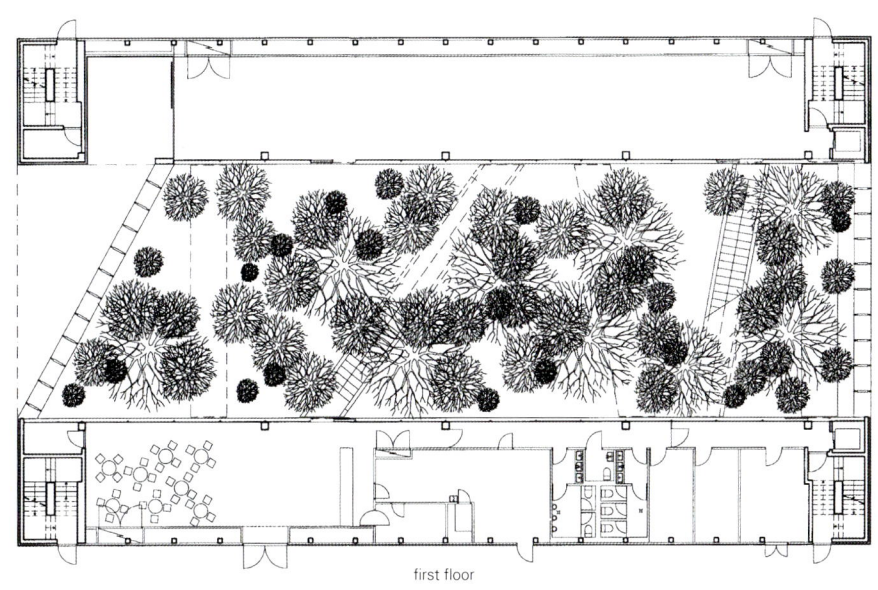
first floor

249

P. S. 1 Dunescape
P. S. 1 둔스케이프

SHoP
SHoP

The Museum of Modern Art and P.S.1 Contemporary Art Center jointly sponsor the MoMA/P.S.1 Young Architects Program, an annual series of competitions that gives emerging architects the opportunity to build projects conceived for P.S.1's facility in Long Island City. SHoP was selected to create the debut project in the series - a dunescape, known as Warm Up 2000, for summer relaxation in P.S.1's outdoor courtyard.

The design provides a variety of ways to enjoy the summer weather - visitors can lounge, socialize, sunbathe, wade in pools, or walk through a spray of water mist to cool off. Five design elements - cabana, beach chair, umbrella, boogie board, and surf - are placed along a continuous wood structure, comprised of over 6000 individual 2" x 2" cedar strips, with a vinyl surface that bends and folds to accommodate various spatial configurations. When the surface is high in the air, it provides shade, when it is lower it provides inclined seating areas. When it is on its side, it becomes a thickened translucent wall, creating individual "cabanas" where visitors may change their clothing. As it twists onto the ground "lifeguard" stands also serve as "dancing" platforms. Water runs along the entire surface collecting in pools throughout the courtyard where the surface touches the ground. A mist garden disperses water throughout the air.

모던 아트 박물관과 P.S.1 현대 미술 센터가 협찬한 모마/P.S.1 신진 건축가 프로그램은 해마다 공모전을 통해 젊은 건축가들에게 롱아일랜드 시의 P.S.1 편의시설을 설계할 기회를 주고 있다. P.S.1 마당에 여름철 휴식을 위한 시설인 웜 업 2000에 SHoP의 프로젝트가 채택되었다.

이 디자인은 여름을 즐기는 여러 가지 방법을 제공하고 있다. 방문객들은 휴식, 사교, 일광욕, 헤엄치기, 물을 뿜는 분수 속 걸어다니기 등을 즐길 수 있다. 다섯 가지 디자인 요소들 - 탈의실, 비취의자, 양산, 디스코 무대, 파도 - 은 나무 구조체를 따라 위치하고 있다. 이러한 요소들은 각각 2" x 2" 크기에 비닐로 표면처리된 6,000개 이상의 길고 가느다란 삼나무들을 묶고 접에서 다양한 형태로 표현했다.

그것은 천정이 되어 그늘을 만들기도 하고, 낮은 곳은 바닥이 되어 앉을 곳을 제공하기도 한다. 또한, 구조체의 측면은 기댈 수 있는 벽이 되기도 하고 방문객의 옷을 갈아 입을 수 있는 탈의실이 되기도 한다. 땅 위에 서로 꼬여서 사람들을 다치지 않게 하는 역할도 하며 무대가 되기도 한다. 바닥에 깔아 물이 흐르도록 하여 풀장도 만들었고, 안개 정원에서는 공기 중에 물을 분사한다.

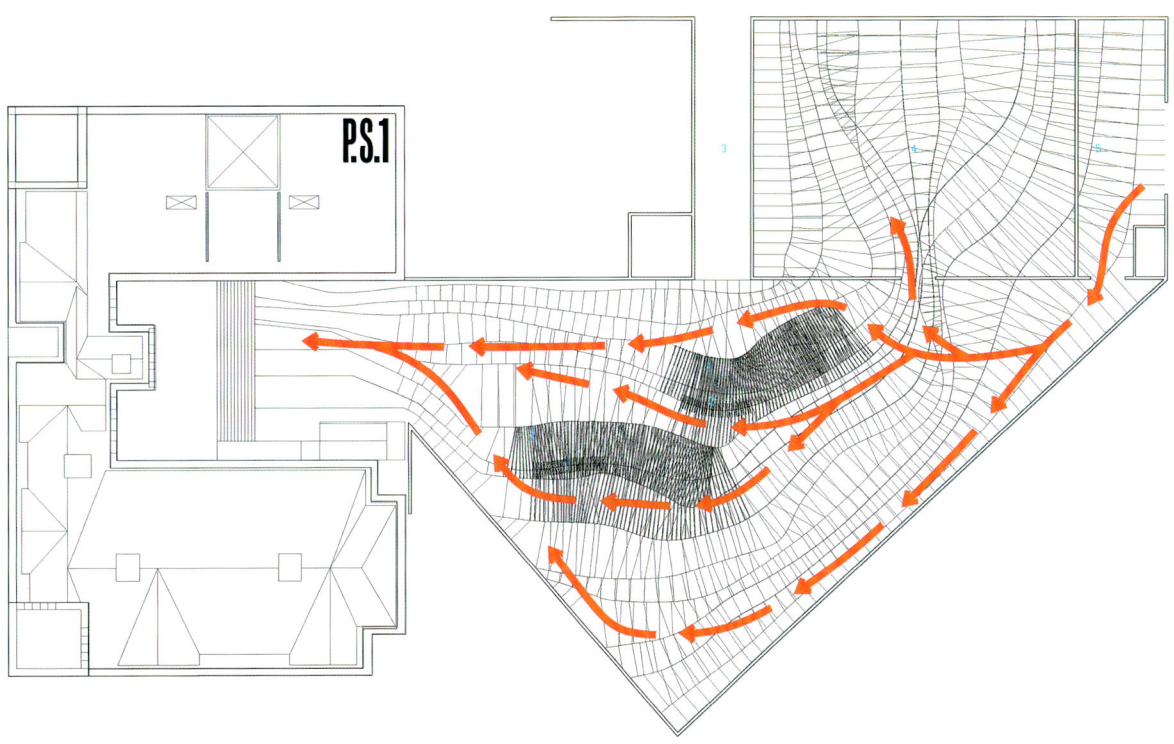

1. cabana 2. reflecting pool 3. service entry 4. DJ Booth DJ 5. entry
floor

Architect: Christopher R. Sharples, William W. Sharples, Coren D. Sharples Kimberly J. Holden, Gregg A. Pasquarelli
Project team: Jonathan Mallie, Richard Garber, Kensuke Watanabe
Fabrication team: Jonathan Mallie (site project manager), Jonathan Baker, Roberto Biaggi, Aaron Campbell, Keith Kaseman, Kris Lawson, Michael Russo, Jamie Palazzolo
Structural engineer: Buro Happold New York
General contractor: SHoP-Sharples Holden Pasquarelli
Client: The Museum of Modern Art P.S.1 Contemporary Art Center
Location: Long Island City, NY
Use: Outdoor installation
Site area: 12,000 sq.ft.
Photograph: David Joseph-p.250, p.252, p.253, p.255, p257
Jonathan Mallie-p.251
SHoP Architects-p.254, p.256

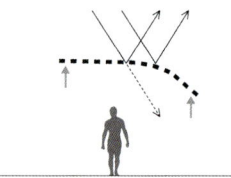 umbrella

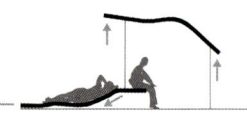 cabana

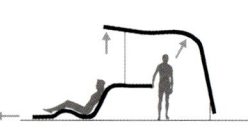 cabana

 beach chair

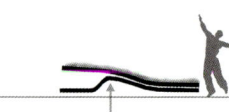 boogie board

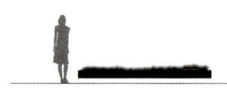 surf

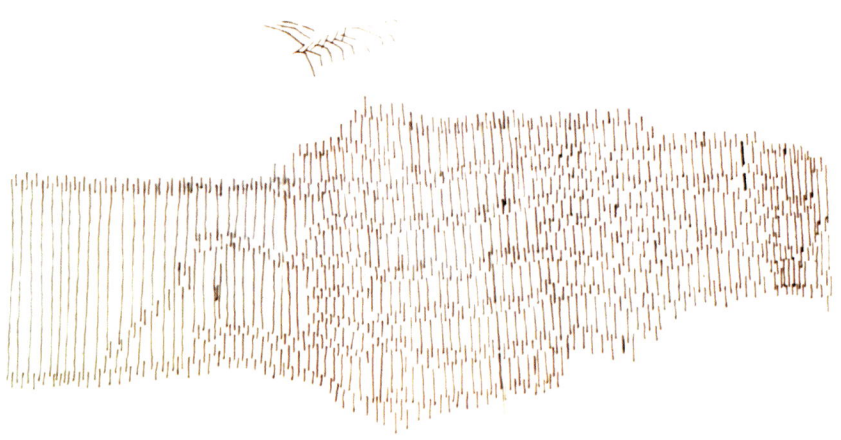

Turkish Pavilion Hannover Expo 2000
하노버 엑스포 터키관

Tabanlıoğlu
타반르오울루

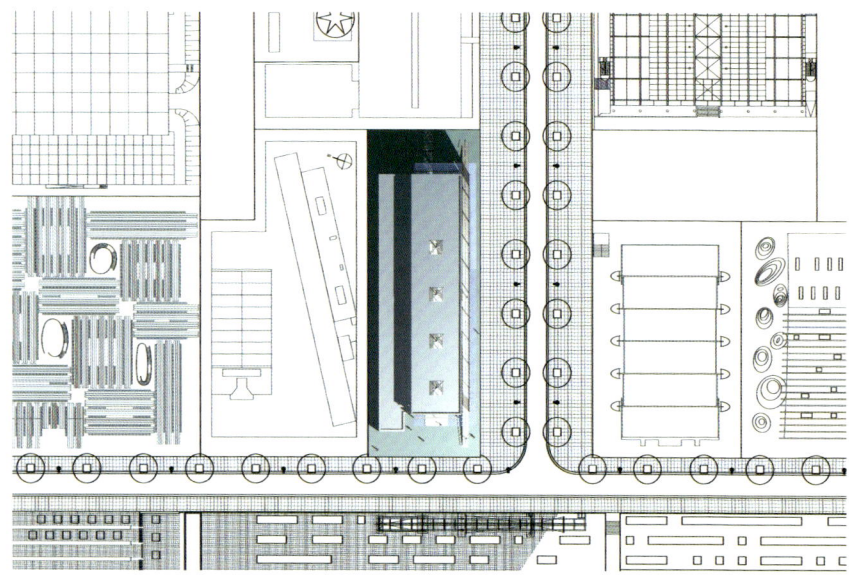

Architect : Tabanlıoğlu Mimarlık
Dipl. Ing. Murat Tabanlıoğlu
Melkan Gürsel Tabanlıoğlu
Architectural project team : Çağman Tepetaş, Hacer Akgün,
Alp Hatipağaoğlu, Ayşegül Kamut
Coordinator : Şeyda Arguner
Technical supervisor : Murat Cengiz
Structural project : Ing. Buro Peter Budde
Mechanical engineer : Ing. GmbH Rühl
Electrical engineer : Ing. GmbH Rühl
Interior design : Tabanlıoğlu Mimarlık
Interior design consultant : Önder Küçükerman
Interior design execution : Arasta, dDf
Exhibition concept design : dDf (Dream Design Factory)
Main contractor : HMB Hallesche Mitteldeutsche Bau AG (Tekfen Group Company)
Landscape consultant : Murat Pilevneli
Lighting consultant : Zeki Kadirbeyoğlu
Sponsor : T.C. İş Bankasi, İMKB
Client : T.C. Turizm Bakanlığı, Ministry Of Tourism
Site area : 2,400 m^2
Bldg. area : 1,943 m^2

Hannover Expo 2000 Turkish Pavilion is a building that is designed in order to be able to reflect both the technological perspective, which is being improved every passing day over the existing cultural riches, and the modern life philosophy to its visitors with a variety of effective messages in a very short time interval.

Today's viewpoint of global construction exposes a philosophy that is based on natural materials, transparent construction, high technology and recyclable structures.

Consequently, this philosophy formed the main design principles of this building, which symbolizes the continuity of the prosperous Anatolian culture together with a contemporary perception. This transparent technological structure, with its 60m x 21m x 13m(L x B x H) dimensions, is designed with a steel construction technique. The Pavilion is made up of five main cubes and has got modernised glass pyramids on its roof.

When Hannover Expo 2000 is examined from the outside, it can be seen that three sides of the building is surrounded by pools that symbolizes Turkey's geographical location. The facade is designed as a transparent glass. The wooden lattice in front of the building, one of the significant elements of Turkish architecture, is modernised with the understanding of contemporary technology and used in front of the glass facade, which was designed for controlling the light.

This lattice was torn apart in order to have the bridge, which takes place behind it, to be perceived from outside and constitutes one of the main design elements. This bridge is the representation of cultural heritage that has been inherited by us from 9,000 years old civilizations of different eras.

Such bridge is also a symbol of a tie between the soil and culture moulded on our land for a variety of time and our civilization approaching to the next millennium. Civilizations, are expressed in various stages in the interior of the pavilion parallel to this bridge. Such stages also reflect different cultural strata. Expressed with symbolised objects, the images reflected onto a huge screen emerging from water by the end of such strata will be one of the major focal points of the visual show presented to the visitors. This huge show, at the same time, will be strengthened and emphasized with the music and natural fragrances from our Anatolia that would be emitted in the hall, causing visitors review any notions that they know or not know regarding real Turkey.

The Turkish sunshine will be expressed by means of special illumination effects created right after entering to the building and the landscaped areas in the interior will be planted with plants and flowers brought from our country to integrate with such a lively atmosphere. The purpose is to express Turkey to the visitors with its varied cultural treasures, natural beauties and the modern face it has achieved as a result of all the foregoing.

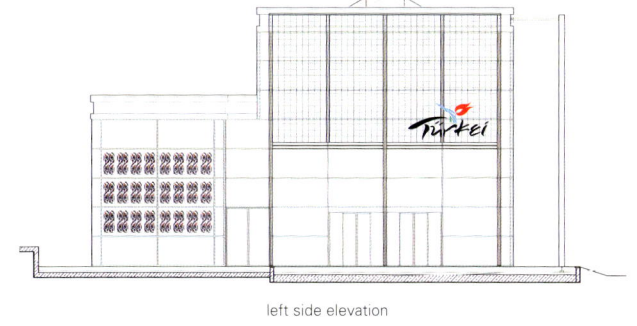

left side elevation

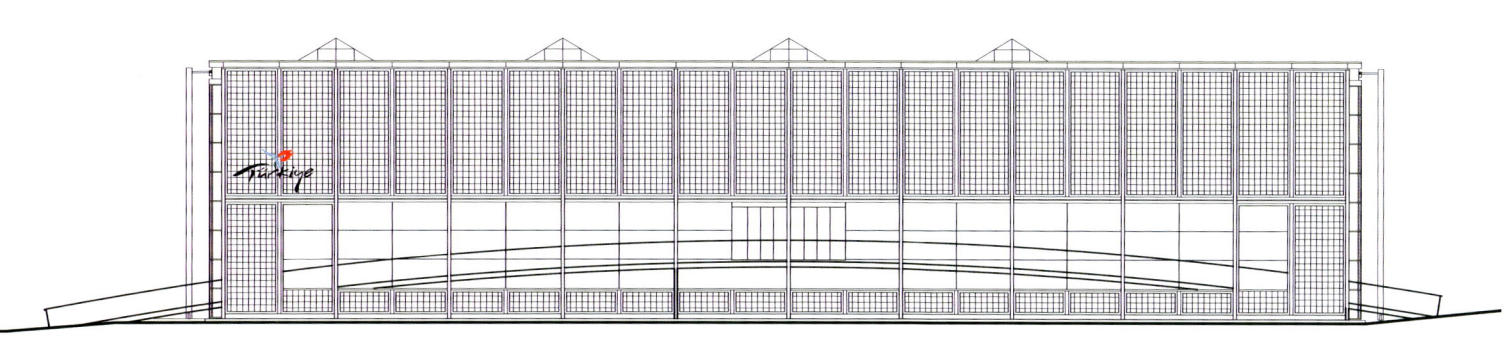

front elevation

하노버 엑스포 터키관은 터키의 풍요로운 전통문화의 토양 위에 나날이 발전하고 있는 첨단기술과 현대적 인생철학을 반영해 엑스포라는 제한된 짧은 시간에 방문객들에게 다양한 메시지들을 효과적으로 보여주는 건물이다.

글로벌 건축물에 대한 오늘날의 관점은 자연 건자재, 투명한 건축, 첨단기술 및 재생 가능한 구조를 토대로 한 친환경적 철학을 보여준다.

터키관은 이러한 친환경 철학을 주제로 과거 번성했던 소아시아 문화와 현대적 감각의 접목을 상징하고 있다. 투명한 구조물 60m x 21m x 13m 에는 강철을 이용한 첨단 기법을 활용했다. 터키관은 다섯 개의 튜브 모양으로 구성했고 현대적인 유리 피라미드 형 지붕이 튜브 위에 자리잡고 있다.

외부에서 보면 건물의 삼면이 물로 둘러싸여 있는데 이는 터키의 지리적 특징을 형상화한 것이다. 파사드는 투명 유리로 처리했다. 건물의 자연채광을 조절하기 위해 이용된 유리 파사드 앞쪽에 목재 격자구조는 터키 건축물의 중요한 요소를 활용해 현대화한 것이다.

격자구조는 두 개로 분할되어 다리로 이어진다. 외부에서도 보이는 다리는 주요한 설계요소 중 하나이다. 이는 구천 년의 장구한 역사를 이어 내려온 터키의 다양한 문화유산을 상징한다.

이 다리는 다양한 시대를 거쳐 터키의 땅에 새겨진 문화와 새 천 년을 맞은 현대 터키 문명 사이의 가교를 상징한다. 다리와 평행을 이루고 있는 전시관의 내부에는 다양한 시대의 다양한 문명들이 표현되어 있다. 특징적인 상징물들을 빌어 표현된 이미지들은 물에서 떠오르는 초대형 스크린에 투영된다. 스크린에 투영되는 이미지들은 관람객들에게 최대의 볼거리다. 아울러 전시관 전역을 은은하게 수놓을 소아시아 시대의 음악과 자연향은 스크린 이미지를 한층 강조한다. 이러한 공간은 관람객들에게 터키의 참모습을 숙고해볼 기회를 제공할 것이다.

터키관에 들어가면 관람객들의 머리 위에 따사로운 터키의 햇살 특수 조명효과 이 내리쬐며 내부의 랜드스케이프 구역에는 터키의 자생목들과 갖가지 화초들이 화사한 분위기를 연출한다. 하노버 엑스포 터키관은 터키의 다양한 문화유산과 자연미, 현대적 얼굴을 관람객들에게 충분히 표현하고 있다.

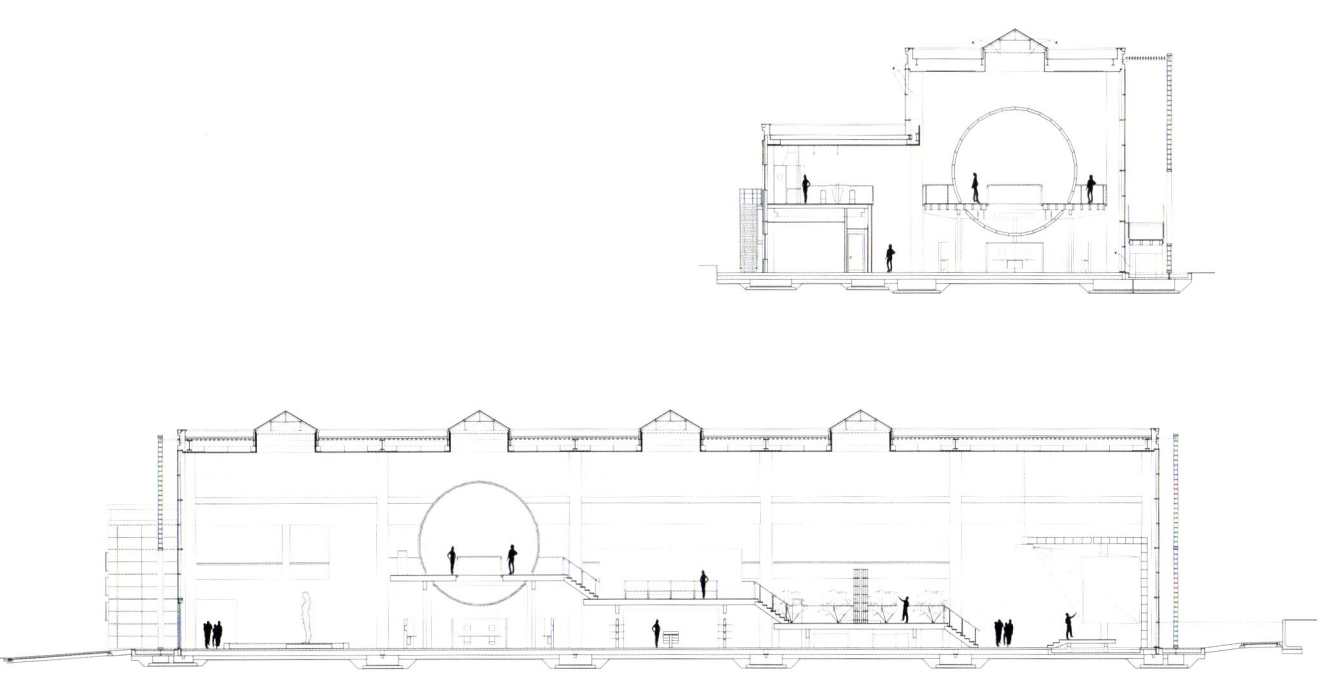

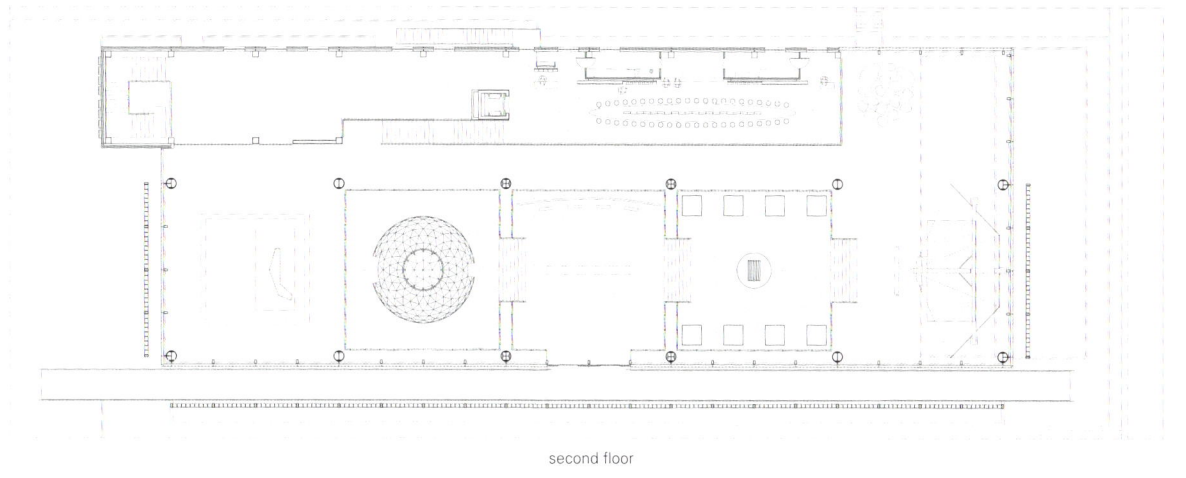

second floor

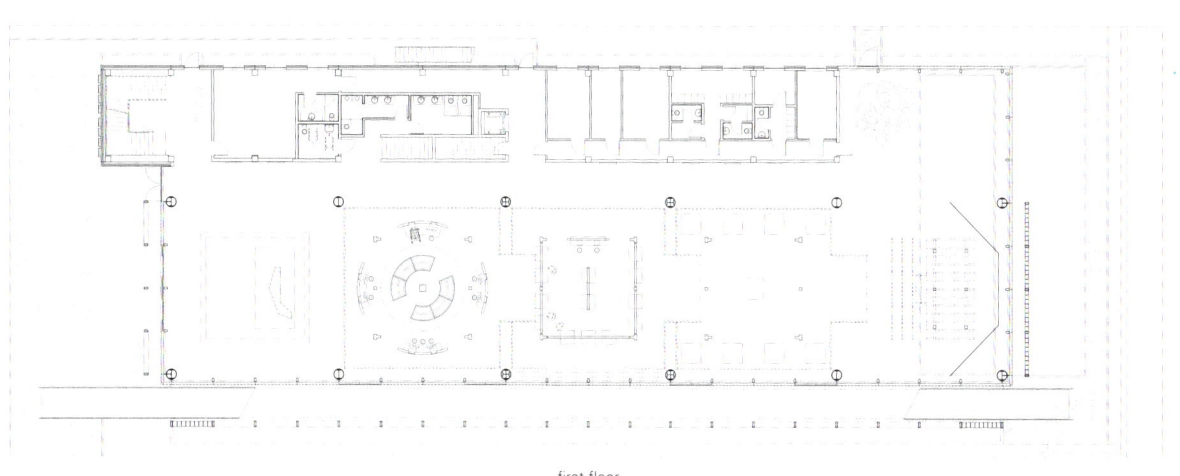

first floor

AREP

Jean-Marie Duthilleul[left] is responsible for all the architectural and development studies for the SNCF and President of AREP. Graduated in Ecole Polytechnique and Graduated engineer of Ecole Nationale des Ponts et Chaussées. From 1980 to 1985 he was teacher of town planning at Ecole d'Architecture of Paris-La-Seine and from 1984 to 1988 lecturer at Ecole Nationale des Ponts et Chaussées. Designed in France with these teams, the renovation projects of Montparnasse and Paris-North stations in Paris, the new stations of Lille-Europe, Aix-en-Provence high speed train stations etc. Abroad, he has realised the project of Nam SeoUl high speed train station and also the Capital Museum of Beijing, Xizhimen Business District in Beijing, TEDA 1 in Tianjin, Shanghai-South station and Porta Susa station in Turin.
Etienne Tricaud[right] is general director of AREP since 1997. Graduated in Ecole Polytechnique and Graduated engineer of Ecole Nationale des Ponts et Chaussées.

장 마리 뒤틸유[left]는 현재 프랑스 국유철도의 건축 및 개발 연구의 책임과 역사 및 도심 통과 구간 개발계획 담당기관의 회장직을 맡고 있다. 프랑스 이공계대학교와 국가기술 전문학교 공학과를 졸업했으며 1980년부터 1985년까지 파리 센 건축학교에서 도시 계획을 가르쳤고, 1984년부터 1988년까지 국가기술 전문학교 토목공학 강사였다. 그는 몽파르나스와 북 파리역사의 리노베이션 프로젝트, 릴과 엑상프로방스 초고속 역사 작업을 통해 프랑스를 디자인 했다. 해외작으로는 남서울 초고속 역사 프로젝트, 베이징 시립 박물관, 베이징의 시지멘 상업지역, 텐진의 TEDA 1, 상하이 남부 역사, 토리노 폴타 수사 역사가 있다.
에티엔느 트리코[right]는 프랑스 이공계대학교와 국가기술전 문학교의 공학과를 졸업했으며, 현재 역사 및 도심 통과 구간 개발계획 담당기관의 본부장이다.

Schmidt, Hammer & Lassen

Morten Schmidt[right] was born in 1956. Graduated from Aarhus School of Architecture in 1982. Member of the Federation of Danish Architects[DAL] and Specialist judge at public Architectural Competitions.
Bjarne Hammer[second right] was born in 1955. Graduated from Aarhus School of Architecture in 1982. Worked at the Jorgen & Kaj Schmidt's Tegnestue and firm of architects Arkitektgruppen for 6 years. Currently, member of the Federation of Danish Architects and director in SHL.
John F. Lassen[second left] was born in 1953. Graduated from Aarhus School of Architecture in 1983. Worked at the Jorgen & Kaj Schmidt's Tegnestue and firm of architects Arkitektgruppen. Mamber of the Federation of Danish Architects and director in SHL.
Kim Holst Jensen[left] was born in 1964. Graduated from Aarhus School of Architecture in 1991. Now, member of the Federation of Danish Architects and director in SHL.
Major projects include the Extension of the Royal Library in Copenhagen, the Culture Centre in Greenland and Aarhus Art Museum and La Rochelle in France.

모튼 스미트[right]는 1956년에 태어났으며 1982년 오르후스 건축학교를 졸업했다. 현재 덴마크 건축협회 DAL 회원이 며 공공건축 설계경기의 전문심사위원을 맡고 있다.
비아네 하머[second right]는 1955년 생으로 오르후스 건축대학교 를 1982년에 졸업했다. 욘 오 카이스미스 타이네스튜어와 아키텍그룹벤에서 6년간 실무를 익혔다. SHL에서 프로젝 트 매니저를 맡고 있으며 덴마크 건축연합의 회원이다.
존 래슨[second left]은 1953년생이며 1983년에 오르후스 건축 학교를 졸업했다. 욘 오 카이스미스 티이네스투어와 아키 텍그룹벤에서 실무를 익혔다. 현재 덴마크 건축연합의 회 원이며 SHL의 주택과 비즈니스 프로젝트의 매니저다.
킴 홀스트 옌슨[left]은 1964년에 태어났다. 1991년 오르후스 건축학교를 졸업했고 현재 SHL에서 디렉터를 맡고 있으 며 덴마크 건축가협회 회원이다.
주요 작품은 코펜하겐 왕립 도서관 증축공사, 그린란드 문 화센터, 오르후스 미술관, 프랑스 라 로셀 등이 있다.

Zvi Hecker

Zvi Hecker was born in Poland, 1931 and spent his teenage years in Samarkand and Krakow. He studied architecture at Krakow Polytechnic, Israeli Institute of Technology in Haifa and painting at the Avni Academy of Art in Tel-Aviv. He set up private practice in 1959, in partnership with Eldar Sharon and Alfred Neumann. He was visiting professor at the Laval Univ., Washington Univ., Iowa State Univ. ect.
Major works include Dubiner Apartment House in Ramat-Gan, Ramot Housing Project in Jerusalem, Art Museum in Palm Springs, USA. Spiral Apartment House in Ramat-Gan, the Jewish Primary School in Berlin, the Palmach Museum of History in Tel-Aviv, the Holocaust Memorial in Vienna. he won architectural competitions, including the Memorial Site in Berlin and the Jewish Cultural Center in Duisburg. He was awarded the German Critic Prize for Architecture.

쯔비 헤커는 1931년 폴란드 생으로 사마르칸트와 크라코 우에서 유년시절을 보냈다. 크라코 폴리테크닉에서 건축을 공부한 후 이스라엘로 이주해 이스라엘공과대학에서 건축 공학을, 애브니 예술학교에서 회화를 전공했다. 1959년부 터 실무를 시작해 엘다 사론, 알프레드 노이먼과 공동작업 한 바 있으며 캐나다 라발대, 텍사스대, 워싱턴대, 아이오 워주립대 등 미국과 유럽 등지의 대학에 출강했다.
두비너 아파트, 예루살렘의 라모트 추가 프로젝트, 미국 팜 스프링스의 미술관, 스파이럴 공동주택, 베를린의 유대 초등학교, 팔마치 역사 박물관, 비엔나의 홀로코스트기념 관 등의 작품이 있으며 베를린의 소실 유태교회 기념공원, 두이스부르그 유대문화센터 등의 설계경기에서 당선되었 다. 1995년 독일 건축비평상을 수상한 바 있다.

Chaix&Morel

Dplg. Architects, Chaix[right] and Morel[left] were born in 1949.
Associated and opened office in 1983. Their architectural achivements have been published in June issue of L'Architecture d'Aujourd'hui in 1986. With their furniture designer, named 'Creators of the year' at the Paris Furniture Fair in 1996.

Ortner & Ortner Architecture

Laurids Ortner[right] was born in 1941, Linz. Studied architecture at Technical University of Vienna, 1959~1965. Worked studio Haus-Rucker-Co in Düsseldorf with Günter Zamp Kelp and Manfred Ortner, 1970~1987. Since 1987 appointment to the State Academy of Art in Düsseldorf as professor of architecture. Member of the Berlin Chamber of Architects since 2001.
Manfred Ortner[left] was born in 1943, Linz. Studied painting and art education at the Academy of Fine Arts in Vienna as well as history at the University of Vienna, 1961~1967. Worked studio Haus-Rucker-Co in Düsseldorf with Laurids Ortner and Günter Zamp Kelp, 1971~1987. Since 1994 Professor for design at the faculty of architecture at the Polytechnic in Potsdam. And member of the Berlin Chamber of Architects.
Founded the Ortner & Ortner Architecture office in Vienna, 1990 and in Berlin, 1994.

I.M.Pei

I. M. Pei was born in China, 1917. Received B. Arch. from Massachusetts Institute of Technology in 1940 and M. Arch. from Harvard Graduate School of Design in 1946. Was an assistant professor in Harvard Graduate School of Design[1945-1948]. Worked for Webb & Knapp, Inc., as a Director of Architecture [1948-1955]. Co-founded Pei Cobb Freed & Partners · I. M. Pei & Partners, New York in 1955.

Jean-Marc Ibos&Myrto Vitart

Jean-Marc Ibos[left] was born in 1957, and Myrto Vitart[right] was born in 1955, they founded their practice in 1989. From 1985 to 1989, they were both associate members of the Jean Nouvel & Associés practice.
Jean-Marc Ibos and Myrto Vitart taught at the Ecole Spéciale d'Architecture in Paris as visiting professors in 1994~1995.

쟝 마끄 이보스[left]는 1957년 생이며, 뮈르또 비따르[right]는 1955년 생으로, 1989년에 사무소를 개소하였다. 이들은 1985년부터 1989년까지 쟝 누벨 사무소에서 실무를 하였으며, 1994부터 1995년까지 파리의 특별건축학교에서 객원교수로 학생들을 가르쳤다.

Kisho Kurokawa

Kisho Kurokawa was born in Nagova, Japan in 1934. Graduated department of Architecture from Kyoto University in 1957. And, received M.Arch. and PhD. Arch. at Graduate School of Architecture of Tokyo University in 1964.
Advisor at Prime Minister of the Republic of Kazakhstan since 2000. And since 2001 senior advisor at the Institute of Urban Planning Board of Capital, Province of Henan, China. Now, membership at Honorary Fellow, Royal Institute of British Architects, Honorary Fellow, American Institute of Architects and International Academy of Architecture, Russia.
Major projects include the Nagoya City Art Museum, the Museum of Modern Art, Wakayama, Saka International Convention Center, Japanese Nursing Association Headquarters, Toyota Stadium, Melbourne Central in Australia, Republic Plaza in Singapore and Kuala Lumpur International Airport in Malaysia.

쿠로카와 키쇼는 1934년 일본 나고야에서 태어났다. 1957년 쿄토대학교 건축과를 졸업하고 1964년 도쿄 대학교 대학원에서 박사 학위를 받았다.
2000년부터 카자흐스탄의 국무총리 고문을 했고, 2001년부터 중국 헤난주 도시계획재단의 수석고문이다. 영국왕립건축가협회의 명예회원이며 미국건축가협회, 러시아 국제건축 아카데미의 회원이다.
주요작으로 나고야 시립미술관, 와카야마 현대미술관, 사카 국제 컨벤션센터, 일본 간호협회본부, 도요타 스타디움, 오스트레일리아 멜버른 센터, 싱가폴 광장, 말레이시아 쿠아라 룸프르 국제공항이 있다.

Heinz Tesar

Heinz Tesar was born in Innsbruck, Austria, in 1939. He studied architecture at the 'Akademie der Bildenden Kunste' in Vienna. In 1973 he opened his own studio in Vienna and another one in Berlin in the year 2000.
Receiving the international 'Tessenow Preis' for his life and work in the year 2000, he was honoured as an unusual architect, who gave the european architecture new impulses with his work. An architect, who defines his positions without any doubts between the late functionalists and postmodernists, with artistic independence, totally rooted in the present, but far away from actual trends and short - lived customs.
Besides building practise he was teaching at the universities of Cornell and Harvard in the USA, in Europe at the ETH in Zurich, the IUAV in Venice and is currently teaching at the Academia di Mendrisio in Switzerland.

하인쯔 테사는 1939년에 오스트리아 인스브룩에서 태어났다. 비엔나에 있는 빌덴 쿤스테 아카데미에서 건축을 공부하고, 1973년에 비엔나에 개인 스튜디오를 열었으며, 2000년에 또 다른 스튜디오를 베를린에 열었다.
2000년에 그의 인생과 작업으로 국제적인 '테세노프 상'을 받으면서 그는 유럽건축에 새로운 충격을 준 '범상치 않은 건축가' 라는 영광을 얻었다. 포스트모더니스트들과 후기 기능주의자들 사이에서 그의 위치를 뚜렷이 정의할 수 있는 예술적 독립성을 가진 건축가로 오늘날 적극적인 지지를 받고 있는데, 그의 작업은 현실의 경향, 일시적인 습관과는 거리가 멀다.
코넬과 하버드 대학교, 취리히 공과대학, 베니스 건축대학교에서 건축 실습을 가르쳤으며, 현재 스위스 멘드리시오 아카데미에서 지도하고 있다.

Wood Marsh Architecture

Roger Wood[left] and Randal Marsh[right] have been in private practice since 1983 after extensive experience in architectural offices in Melbourne and graduating from the Royal Melbourne Institute of Technology are both Directors of Wood Marsh Pty Ltd. Architecture.
During this period the company's design work has won twenty-seven RAIA Awards for Excellence including the 1998 Victorian Architectural Medal and the National Walter Burley Griffin Award. Wood Marsh Architecture has also attracted international recognition through publication and lectures, and been shown in numerous exhibitions of architecture and furniture design.
Academic contributions have been made nationally and internationally through public lectures and tutoring at Universities and Institutes. In addition to a full range of architecture Wood Marsh Architecture have broad experience in interiors, urban design and have designed sets and installations for fashion parades, visual art exhibitions and performance works.

로저 우드[left]와 랜달 마쉬[right]는 멜버른 왕립기술학교를 졸업하고 1983년까지 멜버른에 있는 건축사무실에서 많은 경험을 얻은 후, 우드마쉬 건축사사무소를 개소하여 운영하고 있다.
이 기간 동안 우수한 안목을 위한 27 RAIA상 등 다수의 상을 받았다. 또한 두 사람은 현재 출판과 강연을 통해 국제적으로 인정을 받고 있으며, 다수의 건축전시회 각구나 자인을 선보이고 있다.
여러 대학에서 강연을 하면서 국내외 학계에 많은 공헌을 했다. 또한 도시디자인에서 인테리어 분야까지 폭넓은 경험을 가지고 패션쇼를 위한 세트와 설치디자인, 비쥬얼 아트 전시 등의 다양한 작업을 했다.

Juhani Pallasmaa

Juhani Pallasmaa was born in 1936 in Hameenlinna. Graduated department of Architecture from Helsinki University of Technology and received Master of Science in Architecture in 1966. Engaged in architectural, product, exhibition and graphic design as well as town planning since 1963. Running his architectural office, Juhani Pallasmaa Architects over the past two decades. Served as Dean of the department of Architecture at the Helsinki University of Technology and Professor of Basics in Architecture[1991-1997]. Instructed as the Visiting Professor at Yale University and served as the Director of the Museum of Finnish Architecture.
The author of numerous books and exhibition catalogues, his current book entitled "The Architecture of Image : Existential Space in Cinema" was published in 2001.
Major projects included renovations of the Finnish Institute, Rovaniemi Art Museum and Sámi Museum and Northern Napland Visitor Center.

주하니 팔라스마는 1936년 해민린나에서 태어났다. 헬싱키 공과대학 건축학부를 졸업하고 1966년에 건축학 석사 학위를 받았다. 1963년부터 건축, 시공, 전시, 그래픽디자인 뿐 아니라 도시계획까지 다양한 작업을 해왔. 현재 20년 이상 개인사무실인 주하니 팔라스마 건축사무소를 운영하고 있다. 1991년부터 1997년까지 헬싱키 공과대학 건축학부 학장과 교수를 역임했으며, 예일 대학교 객원교수와 핀란드 건축박물관장으로 활동했다.
수많은 전시 카탈로그와 도서의 저자이기도 하며 2001년에는 '이미지로서의 건축 영화에서 확장된 공간'이라는 제목의 책을 발간하기도 했다.
주요작으로 핀란드 대학 리노베이션, 로바니에미 미술관, 사미박물관과 방문객 센터 등이 있다.

Bianchini & Lusiardi associates

Riccardo Bianchini[left] was born in 1966. Graduated in Architecture at the Genoa University. Taught at the Milan University from 1995 to 2003, initially with Achille Castiglioni and then as professor-in-charge.
Federica Lusiardi[right] was born in 1965. Graduated in Architecture at the Milan University. Initially worked mainly on the planning of public spaces and subsequently also on digital media design.
They established the Bianchini&Lusiardi associati in 2000 and actually based in Cremona, Italy ; its activity deals with many branches of architecture from building restoration to interior design for museums and temporary exhibitions.
In October 2002 the firm won the grand prize in the 2nd DBEW international design competition in SeoUl ; in 2003 won the first prize in RIBA's Cleopatra's Kiosk competition in London ; it has been also finalist in the competition for interior design of the Museum of contemporary art in Rome, in 2004.

리까르도 비안키니[left]는 1966년 출생으로 제노바 대학교에서 건축공부를 했다. 1995년부터 2003년까지 밀라노 대학교에서 처음에는 아킬레 카스킬리오니와 함께 담당교수로서 가르쳤다.
페데리카 루시아디[right]는 1965년 생으로 밀라노 대학교를 졸업했다. 처음에는 공공 공간 계획에 주력을 했으며 이후 디지털 미디어 디자인 작업을 했다.
두 사람은 2000년에 비안키니 & 루시아르디 사무실을 개소했으며 이탈리아 크레모나를 주무대로 박물관과 전시회를 위한 건물복원에서부터 인테리어를 주요분야로 다루고 있다.
2002년 10월 서울에서 열린 제 2회 DBEW국제인테리어 디자인 공모전에서 주요수상, 2003년 영국왕립건축가협회에서 개최된 클레오파트라의 키오스크 공모전에서 1등 수상, 2004년 로마 현대미술관 인테리어 디자인 공모전에서 최종명단에 들어갔다.

Michael Hopkins

Michael Hopkins was born in 1935 in British. Received Diploma from the Architectural Association in 1964.
Senior Partner of Hopkins Architects, which he co-founded in 1976. Awarded a CBE and knighted for services to architecture, and won the RIBA Gold Medal for Architecture in 1994 [with Patty Hopkins]. He is also a Royal Academican in 1992, a Trustee of the British Museum, 1992~2004 and a past President of the Architectural Association, 1997~1999.
Major works include Manchester Art Gellery, National College for School Leadership and Jubilee Campus, University of Nottingham.

마이클 홉킨스는 1935년 영국 생으로 1964년 AA스쿨을 졸업했다.
1976년에 홉킨스 건축사무소를 설립했다. 1994년 건축가를 위한 RIBA 금메달을 받았다. 1992년에 왕립 아카데미 회원이었고, 1997년부터 3년간 AA학장을 역임했으며 1992년부터 현재까지 영국박물관 이사를 맡고 있다. 주요작으로 노팅엄 대학교 주빌레 캠퍼스와 국립교육대학, 맨체스터 아트 갤러리 등이 있다.

Atelier Kempe Thill

André Kempe [left] was born in Freiberg, 1968. And Oliver Thill [right] was born in Karl Marx Stadt, 1971. They graduated Technische University in Dresden in 1996. Studied the urban in Paris, 1993 and Tokyo, 1994. Established Atelier Kempe Thill in Rotterdam, 2000. Since 1999, visiting lectureship TU Delft, Academie van Bouwkunst Arnhem and Academie van Bouwkunst Rotterdam.
Atelier Kempe Thill takes this modern paradox very consciously as the point of departure of the work. The office is by this able to create structures that are neutral and economic as well as being enjoyable and innovative.
Worked various commissions for housing and public buildings in the Netherlands and Germany. The working field of the office belongs within the realms of urban and landscape planning, public building and housing interiors and exhibition design but also research, scenarios, education and publishing. Major projects include Rowhouse in Roosendaal, Town houses in Osdorp Amsterdam and Youth Hostel in Prora.

안드레 쳄페[left]는 1968년 프라이베르크에서 태어났으며, 올리베 틸[right]은 카를마르크스슈타트에서 1971년에 태어났다. 두 사람은 드레스덴 공과대학을 1996년에 졸업했고 파리[93]와 동경[94]에서 도시에 관해 공부했다. 2000년에 로테르담에 아뜰리에 켐페 틸을 세웠다. 1999년부터 델프트 공과대학, 아른헴 건축 아카데미와 로테르담 건축 아카데미에서 강의를 했다.
아뜰리에 켐페 틸은 작업의 시작에서부터 중립적이고 경제적일 뿐 아니라 유쾌하고 혁신적인 새로운 구조물에 의해 현대적 파라독스를 만든다. 네덜란드와 독일에서 다양한 주택과 공공건물 프로젝트를 했다. 사무실의 작업 영역은 도시 랜드스케이프계획, 공공건축, 주거 인테리어, 전시디자인분야뿐 아니라 연구, 시나리오, 교육, 출판도 한다. 주요작으로 루젠다에 있는 로우하우스, 암스테르담 타운 하우스, 프로라 유스호스텔 등이 있다.

GAD

GAD [Global Architectural Development] is a New York Based company, whose partners are Gokhan Avcioglu [left] and Ozlem Ercil [right], which performs architectural practice and research. Founded as ga architects by Gokhan Avcioglu in Istanbul, in 1994 the firm was renamed GAD in 2000 and moved its office to New York in 2003. GAD's projects range from residential and commercial buildings, urban design and gallery installations to furniture and products. These include an expedition center in a forest in Istanbul, the renovation of former Sultan's wife summer palace in Istanbul, a contemporary Asian fusion restaurant, a factory headquarters in Izmit, skyscraper project for any world city and number of residences in Turkey and USA.

GAD는 고한 아브지오그루[left]와 오즈렘 에르실[right]이 함께 건축 작업과 연구를 하는 건축사무소이다. 1994년 고한 아브지오그루가 이스탄불에 ga아키텍츠를 설립하였고, 2000년에 GAD로 명칭을 바꾸었으며 2003년 뉴욕으로 근거지를 옮겼다. GAD의 업무분야는 주택, 상업건물, 도시디자인, 전시장 설계 등이며, 주요작으로는 숲속의 전시장, 황후의 여름별장 리노베이션, 아시아 퓨전 레스토랑, 공장 본사, 마천루 계획안 등이 있다.

Gluckman Mayner Architects

Richard Gluckman, FAIA, received his BA in 1970 and MA in 1971 from Syracuse University. Completed a wide range of institutional, commercial and residential projects, with a special emphasis on art- and exhibition-related buildings. In addition to his international practice, Richard Gluckman has been a visiting critic at Harvard, Syracuse, and Parsons and has sat on numerous academic and professional juries. Peer Reviewer for the Government Services Administration.
David Mayner, who has worked with Richard Gluckman since 1980, was named Partner in 1998 following the completion of the widely-acclaimed expansion project for the Whitney Museum of American Art. Received a Master of Architecture in1976 from M.I.T. in Cambridge, Massachusetts, and also holds a Bachelor of Science in Mechanical Engineering and a Bachelor of Science in Art & Design in 1973 from M.I.T.

SARC Architects

Antti-Matti Siikala[좌] was born in 1964 in Turku, Finland. Received a Master of Science in Architecture from Helsinki University of Technology in 1993. Worked at Juhani Pallasmaa Architects[1986-1995], Narjus-Siikala Architects serving as Partner[1990-1995]. Established the own office, SARC Architects in 1995. Professor at the University of Arts and Crafts, Helsinki, Interior and furniture design department since 2002.
Sarlotta Narjus[우] was born in 1966 in Turku, Finland. Received a Master of Science in Architecture from Helsinki University of Technology in 1996. Worked at Heikkinen-Komonen Architects[1989-1990] and Narjus-Siikala Architects serving as Partner[1990-1995]. Since1998 working at SARC Architects serving as Partner. Assistant professor at Helsinki University from 1995 to 2000.
Major projects include Finnmap Consulting Office Building, Swiss Nationl Gallery, Media-Art School and Expo 2000 Hannover Finnish Pavilion.

Schweger + Partner

Peter P. Schweger was born in 1935. Studied architecture at Technical University, Budapest and Swiss Federal Technical University, Zurich. Co-founded Architekten Graaf - Schweger in 1968 with Heinz Graaf and renamed the office in Architekten Schweger + Partner in 1987.

SHoP

SHoP is a New York design firm with five partners whose education and experience encompass architecture, five arts, structural engineering, finance and business management. Founded in 1996.
William Sharples[left] holds a Bachelor of Architectural Engineering [five year professional degree] from the Pennsylvania State University, and a Master of Architecture from Columbia University[1994]. Worked as Structural and Project Engineer for Clark Construction Group in Bethesda, Maryland. Currently on the faculty at the Parsons School of Design and is a registered architect in the State of New York.
Coren Sharples[second left] holds a Bachelor of Science from the College of Business and Management, University of Maryland, and a Master of Architecture from Columbia University[1994]. Prior to the founding of SHoP, interned at Raphael Vinoly Architects. She is a registered architect in the State of New York.
Christopher Sharples[middle] received Bachelor of Fine Art and Bachelor of History degrees from Dickinson College, and Master of Architecture from Columbia University[1990] where he graduated with honors for excellence in design. He is currently on the faculty of the City College, City University of New York and is an Adjunct Professor of Architecture at Columbia University.
Gregg Pasquarelli[second right] received a Bachelor of Science from the College of Commerce and Finance at Villanova University, and a Master of Architecture from Columbia University[1994]. Following graduation, was a designer, project manager at the office of Greg Lynn FORM. Currently an Adjunct Professor of Architecture at Columbia University.
Kimberly Holden[right] earned a Bachelor of Art in Art History from the University of Vermont and a Master of Architecture from Columbia[1994]. Prior to the founding of SHoP, was project designer at Greg Lynn FORM, New York.

SHoP는 건축디자인, 예술, 구조, 기술, 재정, 경영을 두루 경험한 다섯 명의 파트너들이 1996년 뉴욕에 설립한 건축 디자인 회사다.
윌리엄 샤플러스[left]는 펜실베니아 대학교에서 건축공학사 학위를 받았고, 1994년에 콜롬비아 대학교에서 석사학위를 받았다. 메릴랜드주 베스다에 있는 클릭 컨스트럭션 그룹에서 일했다. 현재 파슨스 디자인스쿨에 있으며 뉴욕 공인 건축가이다.
코렌 샤플러스[left, right]는 마릴랜드 대학교 경영학 학위를 받고 콜롬비아 대학교에서 1994년 건축학 석사학위를 받았다. 라파엘 비뇰리 사무실에서 인턴생활을 했고, 현재 뉴욕에 공인 건축가로 활동 중이다.
크리스토퍼 샤플러스[middle]는 예술학과 사학사 학위를 디킨슨 대학에서 받았고 콜롬비아 대학교에서 1990년에 디자인에 탁월한 인정을 받고 건축학 석사학위를 받았다. 현재 콜롬비아 대학교 건축학과 조교수로 재직 중이다.
그레그 퍼스쿠렐리[left, right]는 빌라노비 대학교에서 무역과 경제학 학위를 받고 콜롬비아 대학교에서 1994년에 건축학 석사학위를 받았다. 그렉 린 폼 사무실에서 디자이너와 프로젝트 매니저로 있었다. 현재 콜롬비아 대학교 건축과 조교수로 재직 중이다.
김버리 홀던[right]은 버만 대학교에서 예술역사학 학위를 받았고 1994년 콜롬비아 대학교에서 건축학 석사학위를 받았다. SHoP에 있기 전에 프로젝트 디자이너로 뉴욕 그렉 린 폼에 재직했다.

Tabanlıoğlu

Murat Tabanlıoğlu[right] was born in 1960. Graduated Department of Architectyre from Vienna Technical University in 1992. Invited lecturer at Uiliz Technical University[1999-2003], Uludağ University[2000-2001] and Istanbul Technical University[2003-2004]. Currently, chamber of Architects, section of UIA in Turkey and member of Foundation for the Preservation of Turkish Monuments and Environment.
Melkan Gürsel Tabanlıoğlu[left] was born in 1969. Studied Architecture at Istanbul Technical University from 1988 to 1993 and received M.Arch. at Politechnical University of Metropolitan Catalonia. Tougth at Bahçeşehir University Department of Architecture from 2002 to 2003 and Istanbul Technical University from 2003 to 204.
Major projects include Warehouse Modern Museum, Hotel Rixos Livertas, Bursa Celik Palas Hotel, Carrefour SA Shopping Center and Kiew Train Station & Hotel.

무랏 타반르오울루[right]는 1960년 생으로 1992년 비엔나 공과대학 건축과를 졸업했다. 일드즈 공과대학[1999-2003], 울루다으 대학교[2000-2001], 이스탄불 공과대학[2003-2004]에서 초청 강사로 활동했다. 현재 국제건축가연맹 터키지사 회원이며 터키 유적, 환경 보존재단 회원이다.
멜칸 귀르셀 타반르오울루[left]는 1969년 생으로 이스탄불 공과대학에서 건축공부를 1988년부터 6년간 했고, 메트로폴리탄 캘리포니아 폴리테크니칼 대학교에서 건축학 석사 학위를 받았다.
바흐체셰히르 대학교 건축과[2002-2003]와 이스탄불 공과대학[2003-2004]에서 학생들을 가르쳤다.
주요작으로 웨어하우스 현대 박물관, 부르사 첼릭 팰리스 호텔, 큐역사와 호텔 등이 있다.